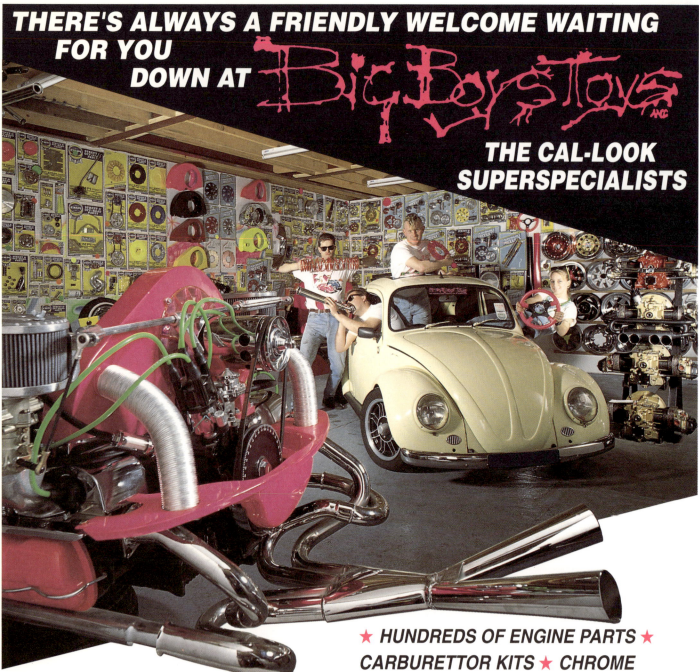

# VW BEETLE in MOTORSPORT

# BEETLE in MOTORSPORT

## The Illustrated History 1940s to 1990s

Peter Noad

Published in Great Britain by
Windrow & Greene Ltd
5 Gerrard Street
London W1V 7LJ

© Peter Noad, 1992

All rights reserved. No part of this publication may be reproduced or transmitted in any form or by any means, electronic or mechanical, including photocopy, recording, or in any information storage and retrieval system, without the prior written permission of the publishers.

A C.I.P. catalogue record for this book is available from the British Library.

ISBN 1 87200 438 5

ghk DESIGN, Chiswick, London

Advertising Sales:
Boland Advertising & Publishing

Printed in Singapore

*Front cover illustrations*
***Left:*** *Jason Collins competing on the VW Owners Club Clee Hills classic trial in his Baja Beetle.*
***Top right:*** *Luke Theochari's 'Moody', built by Terry's Beetle Services, is the first Beetle (with VW floorpan and standard roof) in Britain to run the quarter-mile in under 10 seconds.*
***Middle right:*** *Peter Harrold was British Autocross Champion in 1972 driving his Autocavan Beetle.*
***Bottom right:*** *Richard Hölzl's road-registered 140mph circuit and drag race Beetle at Eastern Creek, Sydney, Australia. (R. Hölzl)*

*Back cover illustrations*
***Top:*** *John Aitkenhead leads a group of Beetle Cup racers in the Bug Prix at Brands Hatch in 1992.*
***Bottom:*** *Mike Hinde and Dennis Greenslade competing on the Charrington's RAC Historic Rally in a 1957 Okrasa Beetle.*

## THE AUTHOR

**Peter Noad** first wrote about motorsport when he reported on rallies and trials, in which he was also a competitor, for *Motoring News* and *Autosport*. In the 1970s he wrote regularly for *Cars and Car Conversions* and was the author of books on *Tuning VWs* and *How To Start in Autocross and Rallycross*. He also contributed chapters on autocross, sprints, slaloms and autotests to *The Castrol Guide To Motor Sport*.
Peter Noad has written regularly for *VW Motoring* magazine (formerly *Safer Motoring*) since 1966 and contributed for eight years to *Volkswagen Audi Car* magazine.
As a motorsport competitor, Noad won hundreds of awards driving VW Beetles. He won the British Trial and Rally Drivers Association Silver Star Rally Championship and the RAC National Autotest Championship, as well as many regional championships. He was a member of the *Cars and Car Conversions* Team and drove the magazine's Hillman Imp in autocross. He won sprint and slalom championships driving an NSU TT and was a class winner on the Total Economy Rally driving an Audi 80.

# CONTENTS

**Acknowledgements** . . . . . . . . . . . . . . . . . 6

**Introduction** . . . . . . . . . . . . . . . . . . . . 7

**The 1940s:** A Beetle goes to Monte Carlo . . . . . . . . . . . 8

**The 1950s:** Irish drivers lead the way; Beetles conquer Africa and Australia . . . 10

**The 1960s:** Beetles invade the British rally and autocross championships…
and the drag race strips of the USA . . . . . . . . . . . . 28

**The 1970s:** Beetles are champions in European rallycross . . . . . . . . . 56

**The 1980s:** Back to the roots; classic trials and historic rallies . . . . . . . . 86

**The 1990s:** The 150mph Beetles . . . . . . . . . . . . . 112

**Appendix:** For those who want to drive their Beetles in motorsport . . . . . 143

## AUTHOR'S ACKNOWLEDGEMENTS

*My thanks go out to the following for providing information, photographs, facts and figures, and reminiscences:*

*Bill Bengry, Gene Berg, Brian Betteridge, John Brewster, Brian Burrows, Liz and Peter Cox (BTRDA), Michael Davis, Tony Fall, Joachim Fischer (Gute Fahrt), Alan Foster (MCC), Ken Green, Dennis Greenslade, Peter Harrold, Mike Hinde, Graham Hoare, Richard Hölzl (Vintage Vee-Dub Supplies), Paddy Hopkirk, Dave Keat, Paul Kunkel, Phill Lander (Club VW Sydney), Dave Lucas, Laurie Manifold, Phil Matthews (Club VW Sydney), Kurt Meyer, Larry Mooney, Klaus Morhammer (Käfer Motorsport), T.P. O'Connell, Tony Royston (Microgiant), Brigitte Seitz (Volkswagen Motorsport), Keith Seume (Volksworld), Kevin Sherry, Ken Shields, Colin Taylor Productions, Luke Theochari (Terry's Beetle Services), Geoff Thomas (Autocavan), Francis Tuthill, Hans Viertel, Robin Wager (VW Motoring), Simon Woodall (VWOC), and Gerry Woolcott (MCC).*

*I would also like to thank everyone who helped my own Beetle to win, in particular Northway Garage (the V.A.G. dealership in Wembley), Autocavan, and Cars and Car Conversions magazine; and especially my navigators, Mick Hayward, Dennis Crome, Alan Harmer, Tony Pryce, Mike Templeman, Brian Culcheth and Sue Granger. They were all, in the words of songwriter Nanci Griffith, 'My Brave Companions of the Road'.*

*Peter Noad*

*Except where otherwise credited, all photographs are by the author using, in historical sequence, Pentax S1a, Spotmatic, MX and LX cameras.*

*T.P. O'Connell driving his oval-window Beetle to outright victory on the 900-mile Irish Rally in 1956. (Mélodie Nightingale)*

# Introduction

When VW Beetles first took part in rallies and trials, some of the cars they competed against were Morris Minors, Ford Prefects, Jowett Javelins, Hillman Minxes, Standard Vanguards, Triumph Heralds, Renault 750s, Fiat 1100s, Simca Arondes, DKWs and Goggomobils. How many of those are still active in all fields of motorsport today?

In the 1990s Beetles are competing in — and winning — rallycross, autocross, trials, drag racing, autotests, rallies, off-road racing, hill climbs and circuit racing. In some sports, such as trials, classic rallying and autocross, they are only slightly modified from standard. In drag racing and rallycross, Beetles have been developed to 400bhp or more. There are Beetles with four-wheel drive, turbos and 16 valves.

An all-Beetle racing series began in Germany in 1989 and is so successful that it is spreading all over the world. Similar Beetle racing is now taking place, or being planned, in Austria, Britain, France, America and Australia.

In this book I have brought together all the diverse fields of motorsport, linked by the unique charisma of the Volkswagen Beetle and its enthusiasts. It has been impossible to mention everyone who has played a part in the Beetle's motorsport history. To all those who have driven a Beetle and won, but do not find their names and pictures in these pages, apologies and congratulations!

# The 1940s

## *A Beetle goes to Monte Carlo*

Established long before the outbreak of World War Two, the Monte Carlo is probably the best-known of all rallies. The first postwar Monte was held in 1949, when petrol was still rationed and there was still very little motorsport of any kind at all.

Three important elements gave the rally its unique appeal. First, the weather: it was held in January, when snow, ice and/or fog were almost guaranteed. Second, the multiplicity of starting-points: in 1949, competitors could choose to start from Glasgow, Florence, Lisbon, Prague, Oslo, Stockholm or Monte Carlo. Third, the 'regularity' tests, of which more below.

Routes from the seven starting-points converged for a common run from Lyon to Monte Carlo. All covered a total distance of 2,000 miles. The required average speed was between 31 and 40mph, with penalties for both early and late arrivals at controls.

As it was unlikely that a winner would be found from the road section alone, there was a further test over three laps of a 17km circuit in the mountains above Monte Carlo. This was a high-speed regularity test. The object was not only to complete it as fast as possible but also to cover two parts of the course (one much more difficult than the other) in identical times. Marks were deducted for any variation in times.

The 1949 rally was won by a 3½-litre Hotchkiss which was reported to have reached 110mph on the test. Others among the top ten finishers were Allard, Salmson, Delahaye and Bristol.

A Volkswagen driven by G. Goedhard finished in 43rd place. In all, there were 230 entries, of which 166 finished. The VW came sixth in the 1500cc class, behind two Jowett Javelins, a Hillman Minx and two Sunbeam Talbots.

When Goedhard entered his VW in the Monte Carlo Rally, *Volkswagenwerk's* total production had barely reached 20,000 and VWs had certainly not yet gained a reputation for reliability.

Goedhard's rally Beetle would have had cable brakes: hydraulics were only introduced on the De Luxe model in 1950. The gearbox had

*Above:* A 1948 Beetle. Owned by Jim Murray, this was not a race or rally car but was in regular daily use on the road when this picture was taken in the 1980s. The registration number is genuine!

*Right:* Original, and well-used, interior of a 1948 standard-model Beetle.

no syncromesh as this was not introduced until 1952. Maximum power of the standard 1131cc engine was 25bhp at 3300rpm.

The Volkswagen air-cooled, flat four engine, and indeed its torsion bar suspension, gearbox and brakes, had competed in motorsport before the 1949 Monte. The first Porsche (prototype 356) was built in 1948 and won its class in a race at Innsbruck. Porsche raised the power output, initially to 40bhp, and Porsche engines were to become a popular means of improving the Volkswagen's performance in motorsport.

Back in 1939, Ferdinand Porsche had built three 'motorsport Volkswagens' for the Berlin-Rome race. These had fully streamlined bodies with enclosed wheels and a narrow enclosed coupé cockpit, and were powered by 40bhp 1100cc engines. The race never took place, due to the war. One car survived the war and was driven by Otto Mathe in races and rallies in Austria in the late 1940s. Several home-made sports cars with lightweight bodies on VW chassis also appeared. Two examples, with 1100cc engines, were raced by Peter-Max Muller and Kurt Kuhnke at the Hamburger Stadtpark Rennen in August 1947. Muller's car was a well-streamlined roadster, while Kuhnke's 'flying saucer'-style coupé featured fully-enclosed wheels and a detachable windscreen and roof section to permit access to the cockpit.

# The 1950s

*Irish drivers lead the way; Beetles conquer Africa and Australia*

Apart from races and speed hill climbs (and the Monte Carlo Rally), any other form of motorsport in the early 1950s was called a trial. Reliability trials are almost as old as the history of the motor car. The classic Edinburgh Trial organised by the MCC (the initials stand for Motor Cycling Club, although it has always catered for vehicles with two, three and four wheels) was first held in 1904. The objective in a reliability trial is not to determine how fast the driver can cover a course, but to see if he or she can complete it.

Trials often involve climbing slippery hills, known as observed sections because performance depends on the observation of a marshal in seeing if, and at what point, the car ceases forward motion. Nowadays tractors and Land Rovers are employed to recover any competitor who becomes stuck on a trials hill. In the early 1950s, horses were used!

Gradually, timed tests were introduced into trials to determine a winner from those who incurred no fails on the observed sections. Night sections were introduced and navigational skills sometimes required.

Rallies originated as 'touring assemblies' — gatherings of participants from different starting-points — and sometimes included a *concours d'elegance*. Speed tests and driving tests were introduced, and more emphasis placed on navigation. They then became known as trials.

Not all the tests were what we would recognise as autotests today. A driving test on a rally or trial in the 1950s might have included width judging or parking manoeuvres, timed 'pit stop' tests where the crew had to remove and replace a wheel, and a top-gear flexibility test with a marshal in the car to see that the driver did not use the clutch. On the Circuit of Ireland there was the famous braking test, in which cars had to approach at 30mph or more and stop in the shortest distance in a curved channel between marker cones.

Passengers were carried in tests (as indeed they still are in trials) and two or perhaps even

*Above: Outright winner of the first East African Safari Rally in 1953 was Alan Dix, driving a split-window Beetle. Dix was then sales manager of the Cooper Motor Corporation, the African Volkswagen importers, who sponsored all the successful VW entries in the Safari Rally. Dix became head of the British VW importers in 1968. The picture shows Dix (far right) with his battered Beetle. An accident several hundred miles from the finish damaged both the Beetle's suspension and navigator John Larsen's nose, but they carried on to win. The other, undamaged VW finished second. (VW Motoring)*

three navigators were sometimes carried on rallies. There were no service vehicles, so all spare wheels and equipment had to be carried on board, often on a roof rack.

Among 282 cars starting the 1950 Monte Carlo Rally was a Volkswagen from Stockholm driven by Cte Einsiedel. As in the previous year, the rally was won by a Hotchkiss. The VW finished in 27th place. In 1951, when victory fell to a Delahaye, a Volkswagen driven by P. Muller finished 41st.

The first major international motorsport success for Volkswagen came on the Tulip Rally in 1951 when T. Koks, a Dutchman, won the 1100-1500cc class in a Beetle. There were 281 starters and the VW was actually third overall, with the outright winner being a Jaguar XK120. The Tulip was a round of the European Rally Championship and was described as 'very arduous'.

For regular VW appearances in motorsport during the 1950s, we have to look to Ireland. Irish businessman Stephen O'Flaherty started importing Volkswagens in 1951 and then formed a company to assemble the cars in Dublin. In 1953 O'Flaherty was also granted the franchise to sell VWs in Britain and started Volkswagen Motors Ltd in London (although the credit for actually introducing Beetles into England belongs to John Colborne-Baber, who began selling and servicing them in 1948).

Irish trials consisted of what the English call autotests, but they took place on public roads, at T-junctions and crossroads, often with hills, varying surfaces and adjacent stone walls and ditches. The route between tests was marked with coloured dye on the road and, instead of marker cones, they used oil drums, buckets and Wellington boots. The organisation may have been informal but the skill of the drivers became legendary: to this day, the world's best autotest drivers have come mostly from Eire and Ulster, and many started with Volkswagens.

Irish drivers invented the handbrake turn and reverse spin turn (throwing the front) and in later years developed tricks such as steering with the knees while operating gear-lever and handbrake simultaneously. It was probably the Irish who first used limited slip differentials in autotests — not to prevent wheelspin but to ensure that both wheels spun together to facilitate tight 180 and 360 degree turns — and experimented with modifications to steering geometry and brake balance.

Two of the earliest recorded appearances by VWs in motorsport in Ireland were by Joe O'Mahoney, who won the novice award on the Munster Autumn Trial in 1951, and L.M. Murray, who competed on the 1951 Circuit of Ireland in a split-window Beetle. Still called a trial, the Circuit included a timed ascent of Tim Healy Pass, driving tests, and a navigational section where many competitors chose the wrong route and got stuck in a bog.

O'Mahoney, a VW dealer in Cork, regularly drove a Beetle to win the saloon class in trials and was the outright winner of the Munster 20-Hour Trial in 1953 and 1954. He also drove a VW in a 50-mile race at Cork in 1954.

Michael O'Flaherty (son of Stephen) drove a VW on the Circuit of Ireland in 1952 and the RAC Rally in 1953, where he made good times on all the tests.

Main opposition to VWs in the saloon class of trials came from Ford Anglias, Renault 750s and Morris Minors. (It's interesting that in Ireland Morris Minors were called 'Beetles' before the name was given to the VW!)

Despite having only 25bhp, a maximum speed of 63mph, and taking 24 seconds to reach 50mph, the Volkswagen actually had the best performance of this group. The 'high performance' popular saloons were the Standard Vanguard and Hillman Minx.

The VW's advantage was superior traction. This was particularly significant on tests involving changes of direction and stop/start manoeuvres, often on loose surfaces and gradients. In 1952, Brendan O'Hara drove a Porsche-engined VW on the Circuit of Ireland. The advantage of the Porsche engine was not just that it had 40bhp instead of 25bhp, but that it was 1086cc as against 1131cc and therefore ran in the under 1100cc class.

By the mid-50s many Irish trials and rally drivers were using Volkswagens. Arthur Ryan, Heber MacMahon, Declan O'Leary, Tommy Connolly, T.P. O'Connell (always known just as 'T.P.') and Paddy Hopkirk all scored regular saloon class wins driving VWs. There were 13 VWs competing on the Circuit of Ireland in 1954 and 19 the following year.

After some success with a self-built Ford Special, T.P. O'Connell's first event with a Volkswagen was when he drove an 1131cc split-window Beetle on the Circuit of Ireland in 1954. He made fastest times on two tests, but lost marks on navigation and finished 16th overall. T.P. says of the VW: 'It had very different handling characteristics to most road cars of the period. Its oversteering tendency required a different driving technique but, once mastered, the VW was capable of higher than average speeds over all types of roads and tracks, despite its low power and low maximum speed. The VW excelled over rough and loose surfaces. It was very reliable and tough and it was almost unknown for a VW competitor to fail to finish an event. It required no maintenance or adjustment from start to finish and was very economical.'

Reporting on the 1956 Circuit of Ireland, *Autosport* described the downhill test at Derreenacrinnig (near Bantry): 'For sheer excitement the best performance was that of T.P. O'Connell (Volkswagen) who shot down the hill, rounded the bends until reaching the last one, overturned, landed on all four wheels, and completed the test in a remarkably good time'. T.P. recalls that the reason he drove on as quickly as possible was that he knew another car was already coming down the test! He went on to record fastest times on several tests after this incident.

The Irish Motor Racing Club's 900-mile Irish Rally of the same year 'turned a blind eye to RIAC recommendations concerning average speeds over third-class roads'. It was reported that time was too short to allow stopping for food. Results showed 'an overwhelming list of successes for the marque Volkswagen, and a particular triumph for T.P. O' Connell'. T. P.'s other outright wins included the Irish Experts' Trial and the Circuit of Munster, and he tied with Paddy Hopkirk for first place in the 1600cc touring car class on the 1955 Circuit of Ireland.

As well as outright wins on trials in Leinster and Connacht, Declan O'Leary won the 1300cc touring car class on the RAC Rally in 1956 with a VW. Cecil Vard rallied various cars, including Jaguars, but he also drove VWs to numerous class wins in Irish trials from 1954 to 1958. Gar O'Brien and Charlie Gunn drove VWs with considerable success in hill climbs and races as well as trials.

Towards the end of the decade the VW trials and rally king was Kevin Sherry, a VW dealer in Monaghan. In 1959, Sherry won the Circuit of Ireland outright in a Beetle, after causing a sensation the previous year by finishing fourth. Sports cars, especially the works Triumph TR3s, were expected to win, but 1959 saw a resounding VW victory with Beetles taking the top seven places. Joe O'Mahoney was second and Tom Burke third.

Sherry's list of outright wins included the Donegal Rally, the Circuit of Munster (twice), the Circuit of Clare, the Ballymena Traders Trophy Rally, Enniskillen Traders Trophy and countless other overall and class wins, navigated by Seamus deBarra. Sherry won the Hewison Trophy driving tests championship twice.

Kevin had an agreement with T. P. (who lived in Bundoran) that they would not both compete on the same event; they competed in events in different parts of the country, so that they could both win!

Modifications to Sherry's Beetle included raised compression, balanced and shot-peened crankshaft, lightened flywheel, enlarged manifold and ports, and a larger venturi in the otherwise standard carburettor. He also fitted narrower piston rings and (for the tests) removed a dynamo brush.

To improve the turning circle Sherry widened the front axle, and to quicken the steering he shortened the drag link by an inch, which gave only 2½ turns lock to lock. He used crossply tyres on the front and radials on the rear, and says he never had an accident.

The most significant Volkswagen motorsport achievements of the 1950s — probably the most significant of its entire history — were in Africa and Australia. Beetles scored outright victories on the Safari Rally, the Redex Trial, the Ampol Trial and the Mobilgas Rally, events which vied with each other for the title of the world's roughest, toughest and longest reliability trial.

It is said that Volkswagen's total dominance of the Round Australia rallies, taking the first six places on the 10,000-mile Mobilgas Rally in 1957 and first, second, fourth, fifth, eighth and ninth places the following year, caused the demise of the event. Entrants in other makes lost interest and the media became bored with yet another Volkswagen demonstration! But VW had by then established a reputation for ruggedness and reliability which endures, throughout the world, to the present day.

The first Round Australia Rally was the 10,000-mile Redex. After finishing second in class in 1954, Laurie Whitehead and Bob Foreman drove a VW to outright victory in 1955, heading a field of 175 cars (including 42 Holdens and 32 Standard Vanguards). The route included boulder-strewn sections, rocky ravine crossings, fast-flowing creeks, and blinding dust. And kangaroos! Set average speeds were up to 42mph. Regulations required the crews to carry emergency food and water for a week: that was how long they could be stranded in the outback if they broke down.

Redex pulled out of sponsorship in 1956 and with both Ampol and Mobilgas wanting to take over there were two events for several years, amid disputes with the motorsport governing body over which was the 'official' event.

In 1956 the 7,000-mile Ampol Trial was unauthorised and entrants were threatened with the withdrawal of their competition licences. Nevertheless, 113 cars took part, among which were nine VWs. Victory went to a Peugeot 403, with a VW driven by M. Goldsmith coming second. Volkswagen, by virtue of also taking fourth and ninth places, won the team prize.

The 1956 Mobilgas Trial attracted 86 entries, including 22 VWs, plus service technicians from the Volkswagen factory. It was a major VW victory, with Beetles coming in first, third, fourth and sixth. There were 11 VWs in the top 17, and again they won the team prize. The winning drivers were Eddie and Lance Perkins.

In 1957 VWs took first and second places on the Ampol (Jack Witter/Doug Stewart and George Reynolds/Lance Perkins), and totally swept the board on the Mobilgas. Laurie Whitehead, Jack Vaughan, John Hall, Eddie Perkins, Bob Foreman and Greg Cusack took the top six places, all in Volkswagens. Of 52 cars which finished the event, 18 were VWs.

The '58 Mobilgas marked the end of the era. It was another convincing victory for VW, the outright winner being Eddie Perkins, with Greg Cusack second and four more VWs in the top ten. But it was the last rough, tough Round Australia marathon for several years. Ampol changed the format. The rally became less of a car-breaker, using faster, better-surfaced roads in the south east corner of Australia, and it resulted in the first victory for the Australian marque, Holden. Jack Witter's VW tied for second place with a Standard Vanguard.

Although not as long as the Australian events, the East African Safari Rally had a similar reputation as one of the most gruelling challenges in the world of motorsport. First held in 1953, when it was known as the Coronation Safari, it was described as a 2,000-mile flat-out race through mud, dust and jungle.

VW Beetles won the Safari Rally four times, in 1953, 1954, 1957 and 1962, and won the team prize in 1953, 1954, 1955, 1957 and 1958. Alan Dix and Johnny Larsen drove the winning

## THE 1950s

Beetle in 1953 — the first outright win for a VW on a major rally.

Vic Preston and D.P. Marwaha scored the second Safari victory for Volkswagen in 1954, and they were followed by Ron Richardson and Monty Banks in another Beetle. The following year, VW only took second place and a class win (Richardson/Banks again), but in 1957 it was VW one, two and three: Gus Hofmann/Arthur Burton, Pat Townsend/Derek Shepherd and Jim Cardwell/Nick Thomas were the drivers.

Volkswagen maintained a presence in European championship rallies throughout the '50s, figuring in the results of the 1952 Monte Carlo (M. Nathan 2nd in class), 1953 Tulip (T.Koks 2nd in class and 7th overall), 1956 Monte Carlo (W. Levy 2nd in class and 5th overall) and 1958 Monte Carlo (Klinken, 2nd in class). Berndt Jansson and Harry Bengtsson scored class wins with VWs on the Rally of the Midnight Sun and the Viking Rally in Scandinavia in 1956, 1957 and 1958.

Volkswagen started to appear on rallies in England in 1954; there were three on the RAC Rally that year. The first notable results were achieved by Bill Mackintosh who drove his VW to second in class on the 1955 Birmingham Post Rally, 11th overall on the 1955 London Rally and third overall on the 1956 Morecambe Rally, these all being national status events. Mackintosh also gained third in class on the 1956 RAC Rally.

The three leading cars on the Morecambe, Mackintosh's VW and two Triumph TR2s, all lost only one minute on the road. John Waddington, the eventual winner on driving test times in his TR2, had been a minute late, but the VW and the other TR2 had been a minute early! A total of 289 cars competed.

In March 1958, M. Crabtree in a VW won Harrow Car Club's Annual Rally. The roads were snowbound and Peter Noad was navigating in a Jaguar driven by his father. Together with many other cars, they were stuck on a slippery, snowy hill near Guildford. When Noad saw the VW which went on to win the rally drive past, not only with enough grip to get up the hill but able to drive off the road and go through deeper snow to pass the stationary vehicles, he decided that when he bought his first car it would have to be a VW.

The Beetle which Noad acquired the following year was a split-window, left-hand-drive model, with cable brakes and crash gearbox. He was told that it was a 1956 car, but that was only the registration, TLC 116. In fact, it had been made in 1946, but had a later 1192cc engine — plus a Wessex twin carb conversion and Konis. Its first rally was the Sporting Owner Drivers Club Dubonnet in 1959 when Noad, navigated by Alan Harmer, came second overall.

The split-window, cable-braked Beetle took Noad to outright wins on the Farnborough Winter Mixture Rally and Chess Valley Three Counties Rally. Other noteworthy results were seventh overall on the Hagley Welsh 12 Hours and seventh on the Mullard Trophy in the Pennines, both against top crews in Austin-Healeys, Mini-Coopers, hot Ford Anglias and Riley 1.5s. After rallying a cable-braked Beetle in the Welsh Mountains and in Yorkshire, Noad regards anyone who criticises the brakes of an early Golf or Polo as a wimp!

Other VW drivers winning awards in rallies and autotests in England at this time were John Wood, Max Rogers, Robin Stretton, Alan Piggott, Peter Crummack and D.M. Barton. Volkswagens were entering rallies in the USA and Canada, and in 1957 Pete Smith and Harry Shap's VW won Canada's 750-mile Caribou Rally. In 1958 Les Stanley and Les Chelminski, in a Karmann Ghia, came second out of 146 starters in the 1,000-mile Canadian Winter Rally.

VWs also did well on fuel economy rallies. On the Hants & Berks Motor Club's famous Mobilgas Economy Run in 1956, the 1600cc class was won by Bill Bengry who achieved 43.02mpg in his VW. Second was Bob Wyse, the founder, editor and publisher of *Safer Motoring* magazine, which later became *VW Motoring*; and third was another VW, driven by D. Watkin. In 1958 Bengry was again a class winner, when he achieved 49.47mpg. Bengry went on to become one of the famous names of VW rallying in the 1960s.

Autocross came into being in 1953. After one or two prototype events, the first recognisable autocross, a half-mile circuit on grass, was organised by the Sporting Owner Drivers Club and London Motor Club, at the foot of Dunstable Downs, in August 1953. VW honours were taken by J.F. Crawley, who won the 1500cc closed car class.

Other early class winners in autocross driving VWs were R.E. Owen, Donald Mills and Mike Hinde. One report, describing Owen's win at a West Hants & Dorset Car Club autocross in 1954, commented: 'the rear engine made controlled power slides appear easy'. Laurie Manifold started winning in 1958 — he was to be a major force in VW autocrossing for three decades.

Mike Griffin and Roy Vaughan (who formed Cartune Ltd) drove VWs in handicap races at Goodwood in the late '50s. VWs also went racing in Australia, Eddie Perkins achieving the first Beetle race win at the Altona circuit in Victoria in 1955. George Reynolds started racing a Porsche-engined Beetle in 1958 and became Australia's most successful VW racing driver.

Beetles began to feature in 'observed section' trials in England in 1955, with Reg Frolich the first regular winner. Classes in trials (and autocross) were based on engine size and open or closed bodywork, not position of engine. When Frolich's VW beat most of the sports cars on the Falcon Motor Club Guy Fawkes 200 Trial in 1957, it was reported that 'trials provide yet another irrefutable argument in favour of independent rear suspension'.

Major developments to the Beetle introduced during the decade included hydraulic brakes in 1950 (De Luxe model only), a change of wheel size from 3½x16 to 4x15 in 1952, syncromesh on second, third and fourth gears in 1952 (De Luxe only), the single oval rear window in 1953, 30bhp 1192cc engine in 1954, larger rear window in 1957, and anti-roll bar (previously only on Karmann Ghia) in 1959. There were also changes to torsion bars and shock absorbers.

Supercharging was a popular means of improving performance. Judson, Shorrocks and M.A.G. all offered supercharging kits for VWs, and a magazine road test of a Judson-supercharged Beetle in 1957 reported an increase in maximum speed from 70 to 83mph and a 0-60mph improvement from 28 to 17 seconds.

Various twin carb conversions were produced. One, called 'Express', was available in the UK from Rally Equipment Ltd, whose proprietor, Les Needham, is now a director at the RAC Motor Sports Association. Needham drove his VW-Express demonstrator at a Harrow Car Club autocross in 1958 and beat Laurie Manifold.

The most famous name in VW tuning, Oettinger (also known as Okrasa), was actually founded in the 1940s, beginning with a hydraulic conversion for the mechanical brakes. In the 1950s Okrasa was producing state-of-the-art VW tuning components, including special cylinder heads, twin carburettors and special crankshafts.

# Nikki CARBURETTERS
## Simply – THE BEST

**NO DIAPHRAGMS   NO SOLENOIDS**
**NO AUTOCHOKE   NO PROBLEMS!!**

*Nikki* – FOR IMPROVED ECONOMY/PERFORMANCE

*Nikki* – FOR ALMOST EVERY CAR IN THE WORLD

*Nikki* – FOR CONFORMITY WITH <u>ANY</u> MOT EMISSION REQUIREMENTS

*Nikki* – FOR SEEING (AND THEREFORE <u>KNOWING</u>) THE FUEL LEVEL

*Nikki* – FOR CONFIDENCE. 100% SPARES AND TECHNICAL INFORMATION BACK-UP

Remember
*"CARBURATION'S FAR LESS TRICKY ...
WHEN YOU FIT A SUPER NIKKI."*

## Nikki CARBURETTERS

204, Muirhead Avenue,
Liverpool L13 0BA

**Tel:** 051-226 1257/ 256 0366
**Fax:** 051-270 2044

Trade and Export enquiries invited

---

# VW ENTHUSIASTS WIN EVERY MONTH WITH VW motoring

- NEWS
- SPORT
- D-I-Y
- TUNING
- LETTERS
- FEATURES

*Plus* **LOADS OF CARS & PARTS FOR SALE**

Available from your Newsagent or ring

Subscriptions/Circulation: (0778) 393652

Editorial: (0242) 677101   Advertising: (0778) 393313

---

34 St. Oswalds Close
Kettering
Northants.
NN15 5HZ

THE SPECIALIST VW MODEL SUPPLIER OFFERS A VAST RANGE OF MODELS BOTH NEW AND OBSOLETE COLLECTORS ITEMS. DIE CAST, TINPLATE, PLASTIC, RESIN MODELS, CONSTRUCTION KITS, AND OUR OWN RANGE OF CLASSIC VW KITS + PEWTER FINISH MODELS. THE MODEL ILLUSTRATED IS OUR 1:32 SCALE (6" LONG) SPLIT WINDOW BEETLE, AVAILABLE FOR **ONLY £23.99 INC. P&P.**

SEND £1.00 FOR OUR CURRENT CATALOGUE – OVERSEAS ENQUIRIES WELCOME
ACCESS/VISA ORDER HOTLINE
**0536 510779**

---

## TELFORD BEETLE CENTRE

Beetles and Campers
Services ★ Restoring
Established 12 years

Tel: 0952 604483
Trench Telford,
Shropshire

---

## RESTO - SERVICE - SPARES
## The Bidford Beetle Company

**American Imported**
Karmann Ghia Doors,
Deck Lids, Bonnets,
all years, also lots
more spares
**Beetle Doors VGC,
optional extra rear
quartz pop outs, all
years. The parts are
all hand-picked by
Stan of The Bidford
Beetle Company
Phone Stan on
0789 490226**

---

The best range of high quality Beetle conversion kits — Roadsters (2 and 4 seater) Targa Convertibles, Vans, Coupes — from £495

**BODYSHELLS**

**KITS**
Six STYLES of fibreglass dodies to bolt on to a stock Beetle floorplan incl. Sedans, Roadsters, Bajas, Ready Coloured, Deseamed.

**★ CAL-LOOK PARTS ★ ACCESSORIES ★ SPARES**

Send £2.50 cheque or P.O. payable to Wizard VWB for our SUPERCOLOUR INFO 20 pages details + prices of WIZARD Quality Products

WIZARD ROADSTERS
373 BUCKINGHAM AVE
SLOUGH SL1 4LU
TEL: 0753-551555   FAX: 0753 550770

---

## Car Care

**VW/Audi Specialist**

All Mechanical
& Bodywork Undertaken

Customising & Restoration
Work Undertaken

**SHAUN BAKER**
0621 743144

---

## CONTINENTAL AUTOSPARES

**64 Haxby Road, York**

VW spares, repairs & accessories.

Parts (0904) 610286
Workshop (0904) 633060

VW BEETLE IN MOTORSPORT

*Above:* Paddy Hopkirk started driving a split-window VW in trials in Ireland, while he was a student at Dublin University. Here he is shown competing in one of his first events, the Irish Motor Racing Club St Patrick's Day Trial, in 1953. The Beetle's small divided rear window was a handicap when tests involved reversing! (P. Hopkirk)

*Left:* Hopkirk in action on a driving test watched by a large crowd on the Circuit of Ireland in 1953. The Circuit covered 1,000 miles and was still called a 'trial'. Note the roof-mounted spotlamp. Hopkirk had not yet started winning — the fastest VW times that year were made by Cecil Vard, Stephen O'Flaherty, Michael O'Flaherty and Liall Collen. (P. Hopkirk)

*Left:* In May 1953 Hopkirk drove this Beetle to fifth place at the Stepaside Hill Climb. This is a different car from that seen in the previous photographs but still an early split-window model. Note the fluted bumper and absence of quarter window vents. It was at another hill climb, at Cairncastle in 1953, that Hopkirk won his first motorsport trophy. (P. Hopkirk)

## THE 1950s

*Right:* Volkswagens first took part in the Redex Round Australia Trial in 1954. Regarded as the world's toughest test of cars and drivers, the 9,600-mile route was mostly on unsurfaced tracks across deserts, riverbeds and mountains. The Australians were accustomed to large, powerful American-type cars, and nobody expected the Volkswagen to survive the rigours of the outback. On its first attempt in 1954, a VW driven by Laurie Whitehead achieved 13th place, out of 246 starters. The following year Whitehead's VW was the outright winner, and from 1955 to 1958 Volkswagens won five Round Australia rallies. Pictured, and showing the tremendous spectator interest in the event, is a control point at Brisbane Showground in 1954. (Marque Publishing Company)

*Above:* In 1955 Beetles finished first and second overall on the Redex Trial, which had been lengthened to 10,500 miles. Pictured here are Laurie Whitehead and Bob Foreman (left) and Eddie and Lance Perkins with their VWs. One 'horror' section of the 1955 trial included 143 creek crossings in 82 miles. 176 cars started that year, but only 57 finished. Success in the Redex Trial undoubtedly helped Volkswagen sales in Australia: by 1959, the Beetle had taken more than 10% of the Australian passenger car market. (Marque Publishing Company)

*Right:* By 1954 Paddy Hopkirk was regularly winning the saloon class in trials and rallies, his skill at carrying out spin turns and 'throws' earning him the nickname 'Handbrake Hopkirk'. This is the start of a Dublin University rally in 1954, with Hopkirk about to start off in a gleaming new oval-window Beetle. Hopkirk still regards the VW as a very good driving test car, with excellent suspension. Most other cars at that time, he recalls, had a tendency to fall to bits! But, he says, you had to be careful doing reverse throws, where overturning was a distinct possibility. (P. Hopkirk)

VW BEETLE IN MOTORSPORT

***Above:*** *In 1955 Paddy Hopkirk won the Hewison Trophy, the trials/driving tests championship of Ireland. This picture shows Paddy with his Beetle during the MG Car Club Trial of that year. Hopkirk won the Cork 20-Hour Trial — regarded as one of the toughest events in Ireland — and was a class winner on the 1955 Circuit of Ireland in a VW. After that he started driving for the Triumph team and then went on to win numerous successes in International rallies, becoming one of the best-known rally drivers of all time after winning the Monte Carlo Rally in a Mini-Cooper. (P. Hopkirk)*

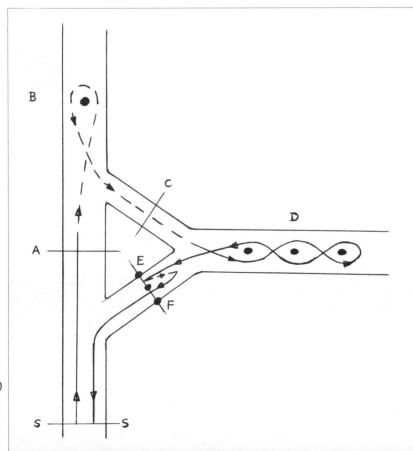

***Left:*** *Diagram showing the type of manoeuvres involved in an Irish trial or English driving test (autotest). In this example cars start from line S, go forward over line A, spin (handbrake turn) and continue in reverse, encircle marker B, reverse over line C, spin (throw the front) and continue forward, in and out of markers D, as shown, continue forward and stop on line E, then reverse off and forward over line F before stopping on line S. If this test was laid out at a road junction, as was customary in Ireland, the road would not be wide enough simply to drive round markers B and D and it would ne necessary to throw the car in a partial spin turn. A test such as this would be completed in about 30 seconds. VW Beetles have been very successful in this type of motorsport from the 1950s through to the present day.*

## THE 1950s

*Above:* Paddy Hopkirk and T.P. O'Connell came first and second in the touring car class in their VWs on the Circuit of Ireland in 1955. 'T.P.' (right of picture) receives his prize money from Stephen O'Flaherty, founder of the company which first started importing VWs into Ireland and Britain, and then assembled VWs in Dublin. Also in the picture are Paddy Hopkirk and his navigator, J. Garvey. (T.P. O'Connell)

*Right:* When T.P. O'Connell won the Irish Rally in his VW in 1956 he beat Triumph TR2s, MGs, Dellows, Ford Anglias, Hillman Minxes, DKWs, and even Paddy Hopkirk in a Standard Ten. Of the test shown here at Skerries, it was reported that O'Connell's VW 'could not hold a candle to the Triumph TR2s in acceleration, but 'T.P.' gave a truly wonderful exhibition of split-second braking to equal fastest time of the day'. (T.P. O'Connell)

*Right:* This VW Special built by T.P. O'Connell in 1959 has a Beetle engine installed in the front, driving a prop-shaft at engine speed to a Beetle gearbox at the rear. The car has a tubular chassis, but all the suspension and mechanical components are from a 1955 Beetle. The wheelbase is 6ft 9in. O'Connell used the car in driving tests and observed section trials, scoring many outright wins including the Irish Experts Trial. (T.P. O'Connell)

# VW BEETLE IN MOTORSPORT

*Above: Not able to afford a car of his own, Mike Hinde started navigating in VWs driven by Dr Robert Ball and Donald Kiff, with whom he won many rallies in Wales and the north-west. Hinde was entered to navigate for Dr Ball on the Severn Valley Motor Club Welsh Rally in 1955, but the doctor was ill, so Mike drove the Beetle himself and won the rally. (Wrekin Photo Services)*

*Left: Mike Hinde competed on the MCC Lands End classic trial for the first time in 1957 and won his class. His Beetle is seen here on the Beggars Roost section near Lynton. Other award-winning VWs that year were driven by Reg Frolich, Ernie Jordan and Eric Jackson. The very first entry by a VW on an MCC classic trial was by I. Cruikshank, who won a second class award on the Exeter trial in 1955. (M. Hinde)*

*Left: A famous name in MCC trials is that of Jack Davis, who competed on motorcycles in the 1920s and 1930s and in an Allard saloon in the early '50s. Jack then bought his first VW (an oval-window De Luxe 1200) which is seen here in 1958 climbing a section known as Stretes, near Ottery St. Mary, on the Exeter Trial. Throughout the next seven years, Davis drove a succession of Beetles (one with a Judson supercharger) in all the MCC classic trials, gaining six first class awards and seven second class. He also won awards on the Falcon Motor Club's March Hare Trial and Guy Fawkes Trial. Davis recalls an incident during the early days of radial tyres when a mismatch of radials on the front and crossply knobbly tyres on the rear caused his Beetle to perform an involuntary 180-degree spin — without using the handbrake. It had the effect of waking up his dozing navigator! (Michael Davis)*

THE 1950s

*Above:* Ireland's leading VW driver in the latter half of the 1950s was Kevin Sherry, who won a great number of trials and rallies. Sherry's first major win was on the Donegal Rally in 1955. He won the Circuit of Munster and the Circuit of Clare, and in 1959 drove a VW to a sensational outright win on the Circuit of Ireland. Sherry won the Hewison Trophy in 1959 and 1960. He also drove his VW to class wins on the Welsh Rally and the RAC Rally, and raced a VW at Kirkistown and Phoenix Park. (K. Sherry)

*Right:* Alan Edmundson won Chester Motor Club's Bernie Rally in 1959. His car was referred to as 'the clockwork Beetle', so as a joke he appeared at the start of the Martini Rally with a large key in the Beetle's engine lid. (Chester Motor Club)

## VW BEETLE IN MOTORSPORT

*Left:* Peter Noad's first car was a split-window Beetle, bought in 1959 and seen here competing in a Harrow Car Club autotest at Denham. Mechanical brakes were a handicap in autotests because the handbrake operated on the same rods and cables as the footbrake. As it braked all four wheels, it was impossible to execute a handbrake turn. Noad drove this Beetle on 29 rallies, winning three outright and finishing in the top six on 12 others. The 1192cc engine had a Wessex twin carb conversion.

*Below:* The 2,000-mile BP Rallies in Australia were perhaps overshadowed by the Mobilgas and Redex events, but the road conditions in Victoria, as seen here, were just as difficult. VWs won the BP Rally in 1958 and every year from 1960 to 1963. (Marque Publishing Company)

THE 1950s

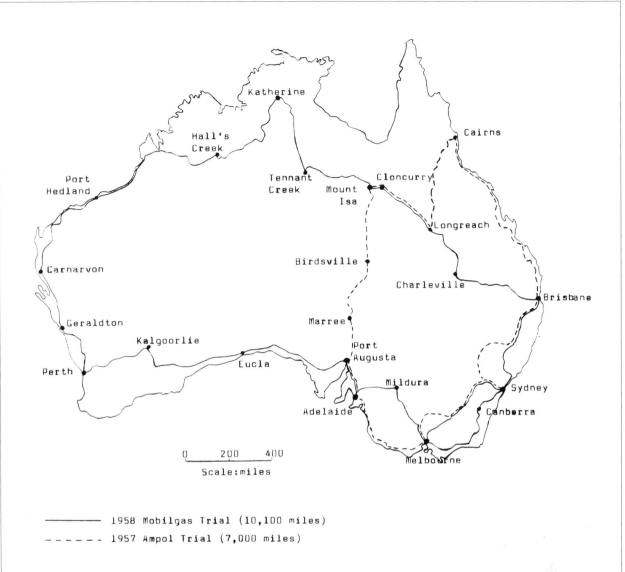

*Above:* Map showing the routes of the 1957 Ampol Trial and the 1958 Mobilgas Trial. Volkswagens won both events outright and there were six VWs in the top ten on the Mobilgas. Both routes ran clockwise starting from Sydney. 'Horror sections' were Mt Isa-Cloncurry and the Birdsville track skirting Simpson desert. Near Perth the route lay over bogs and sandhills, and through water a metre deep.

*Right:* Kevin Sherry competed in the RAC Rally in 1959 and 1960. He was a class winner in 1959. Here his Beetle is seen on the speed test at Brands Hatch in 1960. (L. Thorpe)

# VW BEETLE IN MOTORSPORT

**Above:** This first in a specially commissioned series of drawings by Mélodie Nightingale shows Noel Gleeson in a stop-and-restart test on the Irish Inter-Club trial, a north-versus-south contest between the Ulster Automobile Club and the Irish Motor Racing Club. As early as 1951 Gleeson was driving his 'sawn-off' Volkswagen in trials in Ireland, combining the visibility advantages of an open car with the Beetle's superior rear-engined traction. This trial included observed sections on muddy and grassy hills, timed autotests, and a test which involved coasting down and up hill to stop with a front wheel exactly inside a marked circle.

**Left:** Declan O'Leary's VW on the 1955 Circuit of Ireland Rally. The scene is an up-and-down hill test at the Cod's Head, a remote peninsula south-west of Killarney, surrounded by the Atlantic, on a road not shown on the map. O'Leary teamed up with the VWs of Arthur Ryan and Joe O'Mahoney and they won the touring cars team prize. Fastest time by a touring car on the Cod's Head test was achieved by T.P. O'Connell's VW.

## THE 1950s

**Right:** *One of the earliest 'buggies' was this Irish trials special built by Alf Potter in 1952, which was based on a shortened VW Beetle chassis.*

**Below:** *Handicap races were a feature of club racing in the 1950s. All types of cars took part, including Volkswagens. This scene from a BARC race meeting at Goodwood in March 1953 shows a split-window Beetle on the starting grid, together with a Renault 750, sports-bodied Fiat 1100, several Rileys, Allards and an SS Jaguar.*

**Bottom:** *Very low tyre pressures are needed for a good performance on observed sections on trials. Tyres then have to be pumped up for road sections and timed tests. Operating the pumps here are competitors on the Limerick Motor Club Winter Trial in 1954, Joe O'Mahoney (split-window Beetle) and Bill Young (Dellow). O'Mahoney won the saloon car class.*

# The 1960s

*Beetles invade the British rally and autocross championships… and the drag race strips of the USA.*

This was the Golden Age of road rallying in Britain. It was possible to run highly competitive road events (at night) with straightforward yet challenging navigation. The legal maximum permitted average speed was 30mph, but cars had not yet been developed, nor roads improved, to such an extent as to make the two-minute mile on unclassified roads easy.

The mountains of Wales and The Lake District, the moors and dales of Yorkshire, the intricate network of sunken lanes in Devon, and the muddy farm-tracks of the south-east, all produced scores of classic rallies. There were many national status road rallies, such as the Rally of the Vales, the Bolton, the Bournemouth, the Cat's Eyes, the Plymouth and the Hopper, the prestigious championship run by *Motoring News*, the British Trial and Rally Drivers' Association Championship, and numerous regional championships.

Rallyists have always had abundant ingenuity, and organisers devised various means of 'stretching' the miles. One-minute sections, where stopping at controls carved a significant chunk from the available driving time, measuring distances along near-impassable tracks so that crews were obliged to go the 'long way', and devices such as Targa timing and supplementary sections, were all invented to make rallies still more competitive without resorting to tricky navigation. Gated roads still featured in 1960s rallies, the navigator having to put the maps down, unbelt, leap out of the car, and open the gate. It was essential to have maps marked with the location, and direction of opening, of all gates.

In the mid-'60s, 'selectives' (road sections timed to the second) and special stages (on private roads where a higher target speed was allowed) were introduced. These began to supersede the driving tests which had previously

been included to resolve ties on the normal road sections.

Successful rally cars needed to be reasonably quick, reliable and strong, adequately waterproofed to negotiate fords (always a feature of road rallies in this era), and have lights pointing in all directions, including swivelling spotlamps on the roof and a really powerful reversing lamp.

Small cars had an advantage in overtaking on narrow single-track roads and negotiating 'impossible' hairpin junctions. In the early years, cars such as the Austin A35 and Standard 10 were popular; then came Ford 105E Anglias, Riley 1.5s and Sunbeam Rapiers. On rougher events, the Cortina GT took a lot of beating. Minis were successful, too, but only on surfaces which were not too rough or muddy.

Herefordshire garage owner Bill Bengry was the most famous of the early VW rally

*Top: Kevin Sherry continued to be the most successful Irish trials and rally driver into the early 1960s, winning events practically every weekend. He won the Hewison Trophy driving tests championship for the second time in 1960. Sherry was one of the first drivers to exploit the reverse spin turn or 'front end throw' manoeuvre which enables the car to change direction in a narrow road without stopping. It is equivalent to the forward handbrake turn, but the reverse turn is performed without use of the handbrake. Seen here in a trial in 1961, Sherry demonstrates how the reverse spin can be quite a spectacle when executed 'left hand down' with the driver's weight on the wrong side. (Brian Foley)*
*Above: Mike Hinde regularly competed in the Land's End classic trial and Trengwainton speed hill climb, both held on the same weekend in the early '60s, using his 1200 Beetle in both events. He invariably won first class awards in the trial and in 1962 he gained one of the elusive MCC Triple Awards for unpenalised performances in all three classic trials, the Exeter, Land's End and Derbyshire. Hinde also won his class in the speed hill climb (pictured), only being beaten when Mini-Coopers arrived on the scene. (James Brymer)*

drivers in Britain. He won the RAC National Rally Championship in 1960 and 1961 with a 1200 Beetle, and then drove a VW 1500 Type 3 Notchback to become *Motoring News* champion in 1962.

Bengry started rallying with Simcas, but never did really well due to troubles with engines and navigators. His first success came when he took a standard 1200 Beetle (a demonstrator) and a new navigator (David Skeffington) to victory on the Yorkshire Rally in 1960.

Bengry entered the RAC Rally that year in a 1200 Beetle, finishing seventh overall and second in class, despite cracking the sump and having nine burst tyres. The VW had been one of only three cars without penalty at the mid-rally control in Peebles, the others being Don Morley's MGA and Erik Carlsson's Saab, and Bengry was leading the class of Sunbeam Rapiers and Volvos. The cracked sump was the result of a heavy landing after a violent yump on Monument Hill, the first-ever special stage used on the RAC. Bengry coasted to the end of the stage minus oil and sealed the crack temporarily with mud before repairing it more permanently the next day with Isopon.

The tyre blow-outs were due to the VW suspension 'tucking under' during hard cornering, forcing the tyre away from the rim and trapping the inner tube. Tyres were early Pirelli Cinturas and the wheels at that time were not designed for radials.

Bill's Beetle was fairly standard, modifications consisting of little more than reprofiled valves, enlarged venturi in the otherwise stock carburettor, modified distributor, Konis and front axle supports.

Other VW rally drivers of note in the early '60s included Stephen Clipston, who took fourth place on the Maidstone & Mid-Kent Hopper National and won several southern rallies, and John Dorton, whose Beetle was navigated by Martin Holmes, the now-famous author of many rally books.

Peter Noad probably won more rallies in a Beetle than anyone else in England, taking the Central Southern Championship title in 1964 and 1965, and the BTRDA Silver Star Rally Championship in 1965. During that period, his red Okrasa Beetle scored 22 outright wins including the Plymouth National, the Isle of Wight Rally (twice), the Port Talbot Red Dragon, the Sutton and Cheam Tempest (twice), the Southern Car Club's Scorpio, the Sporting Owner Drivers' Club Dubonnet and the South Bucks Midnight Rally (twice). The Midnight was the first club rally to consist entirely of special stages. Noad's Beetle was a top six finisher on the Morecambe National and the Bournemouth National, and a frequent winner in Wales.

Martin Holmes wrote in *Autosport* in 1964: 'Peter Noad and Mick Hayward scored a victory on the Plymouth National Rally in their faithful old Volkswagen and further strengthened the legend that they are unbeatable whenever the going is rough and slippery.' Reporting on the Tempest Rally for *Autosport* in 1965, Stuart Gray wrote: 'There was no section that was all tarmac — most consisted of up to ten miles of farm tracks, ridgeway and fields. Peter Noad of course revelled in the rough going, bringing yet another win to his VW navigated by Mick Hayward.'

336 BGP didn't quite win *every* rough southern rally. There was an incident on the August Moon Rally in 1963 when the Beetle nose-dived into a grassy hollow on the Berkshire Ridgeway. The impact opened the front bonnet and ejected the spare wheel which rolled away into the darkness. Unable to afford losing the wheel, the crew lost about ten minutes searching for it in the long grass.

Any distortion of the Beetle's front bodywork tilted the headlamps down, making the rooflamp the only source of long-range illumination. The Beetle was good at fording rivers, but one had to remember to turn the heater off, otherwise water entering the air-cooling system would fill the car with steam. There was also the pungent smell of hot mud, baking on the exhaust system, wafting through the heater!

The 1960s is also remembered as the Golden Age of autocross, when almost-standard cars going sideways in the dust and mud made an entertaining spectacle for the crowds and low-cost motor racing for the drivers. Autocross had its heyday in the late '60s when tobacco sponsorship boosted the National Championship, but was subsequently upstaged by rallycross.

When Laurie Manifold won the BTRDA Championship in 1966, he still drove his Beetle to the circuit and his wife used the car to transport their dogs. A Fleet Street journalist, Manifold autocrossed Beetles for nearly three decades, winning more than 200 awards. He achieved outright fastest times (beating all sports cars and specials) on 40 occasions and collected around 30 regional class championships.

In 1960, Manifold's green Beetle had no more than a Minnow-Fish carb, thicker anti-roll bar and lowered suspension. When he transferred these bits to a white 1961 model, he found it did not handle so well (VW had changed to softer torsion bars for '61) and was beaten by another Minnow-carbed VW driven by Lee Atyeo, a west country chicken farmer. Variflo shockers and modified bump stops cured the handling problem, and increased compression and bigger valves provided a slight increase in bhp, but Manifold then found himself beaten by Ken Piper's DKW.

By 1963, Manifold was on his fourth Beetle and spent the massive sum of £169 on modifications. The Minnow carb and suspension mods were as before, but the new demon tweak was wheel spacers. He also fitted some earth-moving tractor tyres, but they were outlawed after one event.

By the time he won the Championship in 1966, the engine had grown to 1470cc (1300 crank and EMPI barrels), the battery had been moved up front, he was running 13-inch front wheels, a Formula Junior gearbox and Spax shockers. Manifold credits Jack Brown of Cambridge Engineering, the original concessionaires for Fish carburettors, with most of the mechanical work.

Volkswagen Motors Ltd (whose previous help had gone no further than advice to buy a Porsche) acknowledged Manifold's success by providing him with a new 1500 Beetle for 1967 and paying for its preparation. There was a snag in that they would only use original Volkswagen or UK-sourced components — no Fish carburettors or American stuff. Engine size was limited to 1620cc (standard oversize Type 3 barrels) and Stromberg carburettors were fitted. With this car, Manifold became class champion in the Player's series in both 1967 and 1968.

Another VW driver to become a British champion during the '60s was Mike Hinde, winner of the BTRDA Production Car Trials Championship in 1961, 1962 and 1963. Mike used 21 different Beetles in trials in two years. He was a motor trader in North Wales, at a time when new VWs were in short supply in England and Wales but were more plentiful in Scotland. Hinde would go by train to Perth, buy a Beetle, bring it back, fit Pirelli Cinturato tyres on the rear and use it on a trial — which he usually won. Then he would sell the car at £50 over list price.

Hinde went on to drive Simcas and Skodas, and in 1966 it was Ken Hoare, from Dorset, who took a Beetle to the top of the BTRDA Trials Championship.

In classic trials there was no national championship: the pinnacle of achievement was an MCC Triple Award for unpenalised performances in all three classics — the Exeter, Lands End and Derbyshire — in one year. (Note: in 1967 the Derbyshire Trial became known as the Edinburgh Trial, for reasons too complex to explain here; it still takes place in Derbyshire and surrounding areas of the Peak District.)

Hinde won a Triple in 1962. Other VW drivers to gain this award were George Edwards in 1960, Ernie Jordan (with a Porsche engine) in 1962, Jack Frost in 1963, D. Hawken (with a Notchback) in 1964, Frank Edkins (with a Judson supercharged 1300) in 1966 and 1969, and Alan Cundy in 1969.

In Australia they raced Beetles in the '60s, most notably in the 500-mile race which originated as the Armstrong 500 at the Phillip Island circuit in Victoria and later moved to the Mount Panorama circuit at Bathurst to become Australia's (and possibly the world's) most famous production car race. In 1962 classes were based on price rather than engine size and cars had to be strictly showroom standard. A VW Beetle driven by George Reynolds and Jim McKeown won the under £900 class. The Beetle completed 162 laps, which was only five laps less than the overall winning Ford Falcon.

In 1963 Bill Ford, who had raced a Hudson Special and was president of the Australian Racing Drivers' Club, persuaded the Sydney Volkswagen distributor Lanock Motors to enter a VW 1200 in the Bathurst race. Bill shared the driving with trials/rally champion Barry Ferguson. Main opposition to the VW in the class came from Minis. Ferguson's duel with the quickest Mini thrilled the 25,000 spectators during the race's final hour. The VW was quicker over the 'mountain' section, but the Mini went ahead each lap on the 'Con-rod' straight. On the last corner of the last lap, Ferguson got the Beetle in front by late braking. The Mini overturned trying to stay with the VW and Ferguson won the class. Two more VWs driven by '62 winners Reynolds/McKeown, and Arthur Andrews/Rocky Tresise, took third and fourth places.

The absolutely standard winning Beetle averaged nearly 60mph during the 500-mile race and needed no attention apart from refuelling. It was reported that the VWs demonstrated superior reliability, better braking, better cornering and less tyre wear than the other cars.

Ferguson won the New South Wales Rally Championship in 1961 and 1962 driving a standard 1200 Beetle, and in 1963 and 1964 in a 1500 Beetle. He came close to winning the Ampol Round Australia Rally in 1964, driving a Type 3 1500S; he finished second and won the team prize together with Doug Stewart and Ray Christie in similar VWs. Christie won the 2,000-mile BP Rally in 1964, driving a 1200 Beetle which had previously been used as a workhorse in Antarctica.

In 1967 Barry Ferguson had a 'works' Beetle fitted with a 1600cc twin-carb engine and limited slip diff. He won the Southern Cross International Rally, competing against drivers of the calibre of Hopkirk, Altonen and Makinen. This was a fast, wet, 2,000-mile rally and Ferguson rates his victory the most enjoyable of all in a motorsport career spanning more than 30 years.

Other rallying highlights of the 1960s were Tommy Fjastad's outright victory on the East African Safari in 1962 (Volkswagen's fourth Safari win) and Harry Kallstrom's second place overall on the RAC Rally in the 1500S (the best result by an air-cooled VW on a European international rally).

In the USA Beetles started to appear in both drag racing and off-road racing in the 1960s, as a consequence of tuning parts becoming available. The originators of VW tuning in the States, now a massive industry, were Gene Berg, Dean Lowry and Joe Vittone. Berg started by increasing the power of his own Beetle — and those of his friends — in the late 1950s, modifying cylinder heads, carburettor and distributor. By 1962 he was installing Corvair pistons and cylinders, enlarging the 1200 VW engine to 1532cc, and he then made manifolds to fit Solex 40 PII carbs on Okrasa heads. Berg built a dragster powered by his modified VW engine and ran low 11-second times on the quarter-mile in 1962. He transported his dragster on a roof-rack on his Beetle and transferred the engine from car to dragster, sometimes running both at the same meeting, swapping the engine between runs, and winning two classes.

About the same time, Joe Vittone formed EMPI, the first company to produce and market VW tuning parts in the USA, with Dean Lowry as head of research and development. EMPI made the first serious inroads into drag racing with a Beetle in 1964, when Lowry drove the EMPI car to record times in the 14-second bracket. Lowry left EMPI to form Deano Dyno Soars, with Berg joining him for a short time as partner. Then in 1969 Berg departed to form Gene Berg Enterprises, a world leader in VW tuning and performance innovations to this day.

By the end of the decade big bore barrels and long stroke cranks were available to enlarge the air-cooled VW engine to more than two litres and twin Weber carbs were fairly commonplace. Dean Lowry and Paul Schley broke into the 11-second bracket with Beetles and Lee Leighton's supercharged VW-engined dragster got under 10 seconds.

Dune Buggies began as recreational/leisure vehicles but moved into motorsport when Bruce Meyers produced a more rugged version — the Meyers Manx — in 1967 and set a new record for the Baja run between Tijuana and La Paz. Baja then became a race and the first winners (in 1967) were Vic Wilson and Ted Mangels, whose VW Meyers Manx did the 920 miles across the desert in 27½ hours.

It was not only Buggies which went off-road racing, but Beetles too. With the front and rear body overhangs chopped off for extra ground clearance, oversize tyres, multiple shock absorbers and upturned exhaust, these became known as 'Baja Bugs'.

In Brazil, in the late 1960s, Emerson and Wilson Fittipaldi constructed a racing Beetle with two VW engines. The two engines were bolted together and installed ahead of the gearbox, making a 3200cc flat-eight-cylinder, mid-engined car, described by Emerson as 'just like a Porsche 917'. Cooling air was ducted in through the roof and two large oil coolers were installed at the front of the car. To the rear of the door pillars the Beetle's chassis was replaced by a tubular space frame and the rear suspension was converted to coil springs. Apparently the Fittipaldi Beetle was capable of lap times faster than a Ford GT40, Lola-Chevrolet and most Porsches!

As well as introducing the Type 3 and the 1500 Beetle during this decade, *Volkswagenwerk* also gave us disc brakes, the Z-bar (to reduce the roll stiffness of the swing axle suspension), the double-jointed rear suspension, and 12-volt electrics.

# ArnieSport
## CHAMPIONS CHOICE
## SINCE 1966

**V.W. SALES • SERVICE • PARTS**

COMPLETE ENGINEERING
FACILITIES
ENGINES FROM MILD TO WILD
AIR OR WATERCOOLED

**0458 42707 or 0935 77312**

SOMERSET, ENGLAND

---

# TERRYS
## BEETLE SERVICES

SHIRLEY GARAGE, SHIRLEY GARDENS,
HANWELL, LONDON W7 3PT

ALL VW MODELS CATERED FOR...ALL BODYWORK AND
MECHANICAL WORK UNDERTAKEN WHETHER CUSTOM OR STOCK......

- Resprays
- Restorations
- Cars bought & sold
- Stock to wild street and full race spec. engines and transmissions
- All welding undertaken
- New & second hand parts available

**Tel: 081-567-3165**

---

## VW DISMANTLERS

Guaranteed secondhand
Beetle spares

Everything from engines
to light lenses.

Fitting, servicing, repairs
and resprays.

Mail order arranged.

Call and save £££s

*Just off the A40*
☎ **Acton (081) 749 9036**

---

## StreetSide
### BEETLES

Race Car Preparation
Cars built to Beetle Cup Spec.
Suspension Modifications
Fast road / competition engines
Disc brake conversions
General servicing / repairs / bodywork

Unit 9, Hawthorne Business Park,
Hawthorne Street, Warrington, WA5 5BX
Tel. 0925 414184

---

**PERSONALITY VOLKSWEAR**

Tees £9.95
Sweats & Jogs £19.95 each + p&p
Hooded Sweats £22.95 + p&p
Also hats, stickers & more

FOR THE FULL STORY SEND
S.A.E. TO:- Unit 1 & 2 Whitecross Trading Estate,
Newquay, Cornwall TR8 4LW
Phone: 0726 861116

---

# FAIL PROOF ®

## VEHICLE ANTI THEFT SYSTEMS

PROTECT YOUR VALUABLE INVESTMENT
WITH THE PROVEN ENGINE IMMOBILISER

THE FAIL PROOF SYSTEM HAS BEEN SUCCESSFULLY PROTECTING
MOTOR VEHICLES FOR OVER TEN YEARS AND ITS UNIQUE
RECORD IN BEATING VEHICLE THEFT IS AVAILABLE FOR YOUR VW
NOW!

USING A UNIQUE, PATENTED OPERATING SYSTEM, EACH FAIL PROOF
INSTALLATION OFFERS THE FOLLOWING BENEFITS:

PASSIVE ARMING (YOU DO NOT HAVE TO "SWITCH IT ON")
NO BATTERY DRAIN WHEN YOUR CAR IS PARKED
INDIVIDUAL, PERSONALISED INSTALLATION
A 100% RECORD IN BEATING CAR THIEVES
ROBUST AND RELIABLE CONSTRUCTION
12 MONTH WARRANTY ON PARTS AND LABOUR
MAY BE USED ALONGSIDE YOUR EXISTING ALARM

FOR MORE INFORMATION ABOUT THIS AWARD WINNING SYSTEM
PLEASE CONTACT:
**MAPLEMAN (UK) LTD.**
45 LAKE RD., AMBLESIDE, CUMBRIA, LA 22 0DF

**TEL/FAX : 05394 32822**

---

## Bugshack
### COACH & MECHANICAL
TEL: 048641 2412

Fast, friendly service to all air-cooled VW's.
*Lowering, dechroming, welding & servicing.
*WIZARD Soft-top conversion *Interiors
*2K Resprays *Engine rebuilds to your
requirements. *Restorations and Customs.

TEL: 048641 2412 *(24hr Answerphone)*
Mobile: 0831 388187
Between Dorking & Guildford

*New & Used spares

---

## BEETLELINK
### VOLKSWAGEN SPECIALIST

Full or Part Restoration • Servicing • MOT
Work • Welding • Mechanical • Resprays &
Bodywork • Cal Look & Custom Work •
Loads of Spares • Cars Bought & Sold •

**Tel: 0252 851590**
Unit 7, Finns Industrial Park
Mill Lane, Crondall, Nr Fleet
Hants GU10 5RP
Only 5 mins from J5 of M3

## THE 1960s

*Right:* Mike Hinde with his Beetle and trophies won in the 1950s and 1960s. (M. Hinde)

*Below:* The Volkswagen Owners' Club of Great Britain ran all types of motorsport events in the 1960s, including a sprint at Brands Hatch. Seen here in 1967 is the standard 1200 Beetle of R. Leeson, about to be overtaken by Peter Colborne-Baber's lhd Okrasa 1300. Colborne-Baber is now Managing Director of Colborne's, the Surrey V.A.G dealership; his father, John, sold the first Volkswagens in the UK in 1952 and founded the Volkswagen Owners' Club. Peter Colborne-Baber made third fastest time in the sprint, behind Norman Higgins' Okrasa 1500 and Peter Noad's Okrasa 1300.

# VW BEETLE IN MOTORSPORT

*Above:* VW owners looking for more motorsport formed the Sporting VW Club, which ran many trials, driving tests and autocross meetings in south-east England. A trials venue regularly used by the SVWC was Canada Heights, not far from Brands Hatch. Competitors had to complete about 30 observed sections, including twisting grassy climbs around gorse-bushes and sandy hills. This is Keith Wilson's 1200 Beetle at a club event in 1963.

*Left:* Production car trials differ from classic car trials not only in the degree of modifications permitted but also in the type of venues and courses. Classics use rugged ancient tracks and roadways; PCT's take place in sloping fields, challenging competitors with tight slippery turns between markers. VWs are successful in both types of trial. A consistent winner in the British Trial and Rally Drivers' Association production car trials during the 1960s was Martin Appleton. Best known for his successes driving a heavily-ballasted Type 3 Variant, Appleton also used a Beetle, as seen here on the Shenstone & District MC round of the BTRDA Championship in 1963.

*Left:* George Reynolds raced a modified Beetle in Australia in the '60s, claiming to have won more than 50 races. His 1300cc Beetle had a top speed of 112mph and a 0-60mph time of 9.3 seconds. Helped by 145 octane fuel and 11 to 1 compression, in 1964 it developed 94bhp. (VW Motoring)

## THE 1960s

**This page:** *George Geshos was another Australian who raced Beetles in the 1960s, competing against big Holdens and Chryslers, as well as Renaults and Minis. The very sideways picture was taken at Oran Park circuit in 1966 when Geshos won his class and finished third overall. The wheel-lifting shot shows him at the same circuit, circa 1968. (G. Geshos)*

# VW BEETLE IN MOTORSPORT

*Left:* In 1963 when Volkswagen introduced the twin-carb version of the Type 3 Notchback (1500S), it was the nearest they ever came to producing a sports saloon — until the Golf GTI. The 1500S scored many rally successes in Scandinavia and Australia and gave VW their best-ever result on the RAC International Rally of Great Britain. In 1963 Harry Kallstrom drove the Scania-Vabis-entered Notchback to second place overall, behind Tom Trana's works Volvo and narrowly ahead of Erik Carlsson's works Saab. (Ray Sargeant)

*Below, left:* The Swedish VW importers entered four 1500S Notchbacks on the 1963 RAC Rally. Behind Kallstrom's class-winning second place overall were Berndt Jansson (13th), Rune Larsson (14th), and Bertil Soderstrom, pictured here at Oulton Park, in 17th place. Kallstrom, Jansson and Soderstrom had all won the Swedish Rally Championship in VWs. Also driving a 1500S on the 1963 RAC Rally was Reggie McSpadden, from Northern Ireland, who finished 35th. (Ray Sargeant)

*Below:* After their success with the 1500S in 1963, Scania-Vabis returned to the RAC Rally the following year with three Okrasa-modified Beetles, driven by Jansson, Larsson and Soderstrom. Jansson won the 1300cc GT class and finished eighth overall, Soderstrom was second in class and 12th overall. In the same year Karl Gudim, from Norway, won the private entrant's award in 16th place overall, driving a 1500S. (Ray Sargeant)

*Above: The twin-carb 1295cc Okrasa engine in the Scania-Vabis rally Beetles of 1964 developed about 65bhp. Okrasa was the only VW engine conversion homologated for International rallies and it put the Beetle into the GT/sports car class. The conversion included a special crankshaft, improving reliability at high revs as well as increasing cubic inches, and twin port heads. Okrasa is an acronym for Oettinger Kraftfahr-technische Spezial-Anstalt.
(Ray Sargeant)*

VW BEETLE IN MOTORSPORT

*Left:* Interior of the Scania-Vabis rally Beetle. Only additional instrumentation is a rev-counter and a clock. Note the specially-fabricated frame for the bucket seat, underseat storage box and wood-rimmed steering wheel.

*Opposite page:* Although it never supplied rallying equipment for the Beetle, the Volkswagen factory did produce a few strengthening or protection items intended for export markets in the more rugged parts of the world, which were often fitted or copied by Beetle rallyists. Some factory items were supplied as kits, others simply as engineering drawings for local manufacturer. The latter included this 'protection plate for front end'.

*Below:* Instrument panel of a privately-entered rally Beetle. This is Frank Andrews' 1300 in which he competed on the RAC Rally in 1966. Smiths instruments replace the VW speedo and fuel gauge, as well as adding a tachometer, ammeter and oil pressure gauge. Halda Speedpilot (on left) was the state-of-the-art average speed calculator for the 1960s rally cars. Split pointers on a clock showed minutes ahead of, or behind, a set average speed. (VW Motoring)

# V – ILLUSTRATIONS FOR THE MANUFACTURE OF REINFORCEMENTS, PROTECTION AND STONEGUARD EQUIPMENT

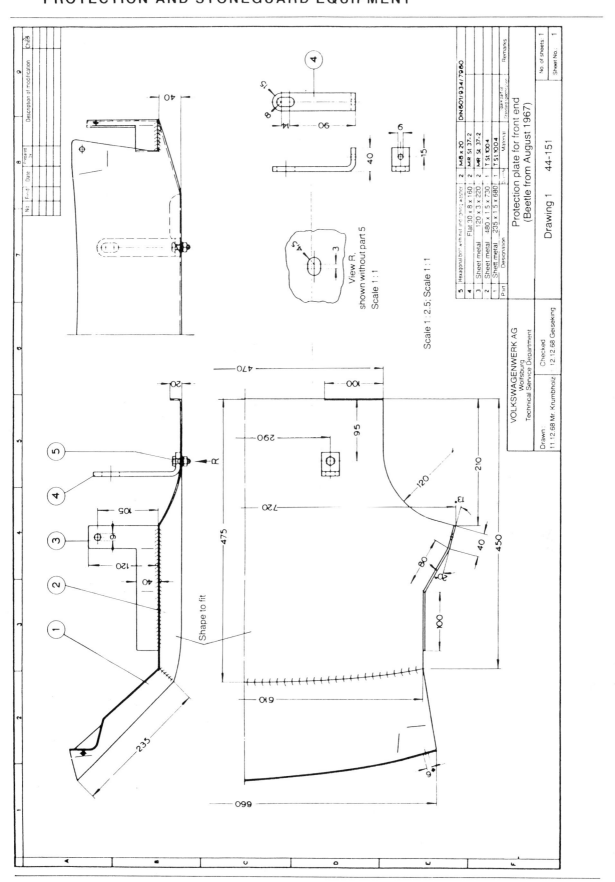

**Left:** Sketch of the front protection plate installation. The plate was bolted to the bumper and the sides of the spare wheel compartment, and brackets attached to the front axle mounting points. This strengthened the front of the bodywork, overcoming the problem of the spare wheel compartment being distorted and pushed back on any impact (from the front or below), and it protected the anti-roll bar which could otherwise become bent by contact with rough ground.

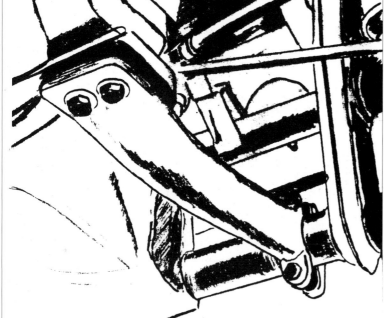

**Left, below:** Another essential item for any Beetle used in motorsport was this front axle support supplied as a kit with VW part number 111 498 001A. This comprises struts that fit between the outer ends of the torsion bar tubes and the front corners of the floor pan. It helps to prevent the torsion bar tubes from bending on any severe impact and minimises flexing of the tubes, with the consequent distortion of steering geometry, that can occur during hard cornering and on rough roads. This picture shows the 1965 arrangement; for earlier models, the support attached to both torsion bar tubes.

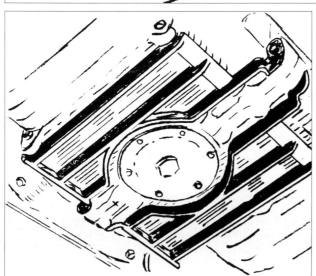

**Below, left:** This is the official 'engine protection grille'. Hardly a sumpguard, it merely protected the crankcase fins and oil strainer plate from minor abrasions when driving slowly over rough ground. For rallying a conventional thick sheet metal plate sumpguard was desirable, but could cause overheating. Many club rallyists fitted the VW towing bar which consisted of two strong skids passing right underneath the engine and served very well as sump protection.

**Below, right:** The Beetle's jacking points can easily be damaged when traversing rough terrain, which causes difficulties when a wheel has to be changed. VW provided a drawing for fabrication of the skid plate, shown in this sketch.

THE 1960s

**Above:** Peter Noad won three rally championships in 1964 and 1965, driving his Okrasa Beetle, 336 BGP, with support from Northway Garage in Wembley. In addition to the 1295cc Okrasa engine, modifications included Koni shock absorbers, front axle supports, Transporter clutch and Porsche wheels. This Beetle competed in 206 motorsport events, won 137 awards, and only failed to finish on eight occasions. Peter became known as 'the king of the white roads' (those marked as unsurfaced on Ordnance Survey maps) and particularly enjoyed mud, bumps and river crossings!

**Above, right:** Peter Noad (right) with navigator Mick Hayward and silverware collected for winning the Isle of Wight Rally (including a trophy presented by the Daily Telegraph) in 1964. Hayward also won rallies, trials, autotests and autocross in his own Beetle, but concentrated more on navigating, where his expertise was always in great demand. Lack of finance was also a factor — navigating was cheaper than driving!

**Right:** Souvenirs of rallying in the 1960s.

# VW BEETLE IN MOTORSPORT

*Above:* In the mid-1960s it was still possible to be competitive in autocross using a road or rally car, although more specialised, lightweight, high-powered, trailer-borne machines were steadily taking over. Peter Noad drove his rally Beetle to a number of class wins in autocross from 1966 to 1968. It was largely due to the success of Beetle's (and later Mini-Coopers) in autocross that the original class structure of the 1950s, based on open or closed bodywork and engine size, was replaced by classes defined by position of engine and driven wheels. Main opposition to VWs in the rear-engine class came from Hillman Imps, plus a few Renault Gordinis and the occasional NSU TT.

*Left:* Laurie Manifold won the BTRDA Autocross Championship in 1966 with this daily-driven Beetle which had covered 50,000 miles and was used by his wife to take their dogs to dog shows. Power (about 75bhp) came from a 1470cc EMPI big-bore conversion and Minnow-Fish carburettor. Handling came from a stiffer anti-roll bar and wheel spacers. Manifold did not actually invent spacers, but was the first to recognise their advantage in autocross.

## THE 1960s

**Right:** *For the 1967 season, Manifold got a new Beetle, sponsorship from the Bishops Stortford VW dealership (known then as Frank Bull's Garage), and a trailer. Engine size was up to 1620cc and carburettors were twin Strombergs. Innovations included a limited slip differential and roof-mounted oil cooler. Manifold was class champion in the Player's No 6 autocross series in 1967 and 1968, recorded several outright fastest times, and picked up numerous regional class championships. Laurie (he's the one with the crash hat and no tie) poses here with sponsor Frank Bull, mechanic Ray Wright, and Fred Thompson of the VW Owners' Club. (F. Scatley)*

**Above:** *The VW factory never officially supported racing or rallying with Beetles but did enter the motorsport arena when Beetle engines and suspension were applied to single-seater racing in Formula Vee. Born in the USA around 1963, Formula Vee was established in Europe in 1966 and became highly successful worldwide as a low-cost, entry-level racing formula. Originally the 1300 Beetle engine was tuned to give 58bhp for Formula Vee and the Beetle's torsion bar suspension, with swing axle and drum brakes, was used. The cars weighed 375kg and had a top speed of around 120mph. Volkswagen Motorsport evolved from the Formula Vee Association (Formel V Europa) and, while it was exclusively concerned with F Vee racing in the 1960s, there was some spin-off of parts and know-how applicable to sporting Beetles. (VW Motorsport)*

**Right:** *Formula Vee racing took off in the late '60s, not only throughout Europe, but also in Australia, South Africa and America. Shown here is a typical Vee race at Daytona. (VW Motorsport)*

VW BEETLE IN MOTORSPORT

**Left:** *Regarded by most experts as the greatest autotest driver of all, Northern Ireland's Robert Woodside often drove Beetles and took over from Kevin Sherry as the driver most capable of achieving outright wins with a saloon car, beating the fastest sports cars, Minis and specials. Here Woodside is driving his 1500 Beetle in an autotest at Omagh in 1967. We're not sure if his driving position — head in the centre of the car — is to get a better view of the marker cone, to get more leverage on the handbrake, or to improve weight distribution! Legend has it that Woodside was capable of some unique and ambidextrous skills with the controls, but perhaps that's just an excuse by mere mortals unable to get within five seconds of his time on an autotest!*

**Above:** *The first Volkswagen driver to win the Westwood Trophy in the British Trials and Rally Drivers' Association Autotests Championship was John Haden in 1966. Haden is pictured here during an autotest at Curborough, organised by the Shenstone Club. The Westwood class was for 'large' saloons, defined as those with a wheelbase greater than 7ft 1inch, which was effectively all saloons other than Minis. The main challenge to Beetles came from MG 1300s, Triumph Vitesses and Ford Escorts.*

**Left:** *Robert Woodside was the saloon class driver in the Northern Ireland team which won the televised Ken Wharton Memorial autotest in 1960, 1961, 1964, 1965, 1966 and 1967. Although no longer televised, the Ken Wharton event, organised by the Hagley & District Car Club, is regarded as the Grand Prix of autotests. Woodside's 1500 Beetle had twin carbs, Type 3 heads, and 1500S (high compression) pistons.*

THE 1960s

*Above:* Eire's autotest team captain, Larry Mooney, is one of the few drivers to have beaten Robert Woodside, which he did at the Ken Wharton event in 1967. Mooney's standard, straight-from-the-showroom 1300 Beetle had crossply tyres, which were an advantage when executing tight spin turns on a grippy surface. Now a special projects manager for V.A.G., Mooney won hundreds of awards in trials, autotests, hill climbs and rallies in Ireland. He was on the front row of the grid in a Beetle in the first-ever race held at Mondello Park in 1968, and he won the Irish Autocross Championship in a 1900cc Beetle sponsored by JCB. Eire never won the Ken Wharton as a team, probably due to a weakness in the sports car class.

*Right:* Cambridge VW dealer Jack Frost won numerous awards with his Beetle. Powered by a 1500cc engine with Minnow carburettor, he competed in autocross and rallies, including the Scottish and RAC internationals in 1965. Using the same car with a 1200cc engine, Frost won first class awards on MCC classic trials. In 1963 he became one of only four drivers to win a prized MCC Triple Award. (VW Motoring)

## VW BEETLE IN MOTORSPORT

*Left:* The Volkswagen Owners' Club (GB) had a regular programme of autotest meetings. Winners during the 1960s included Colin Bishop, Bob Lockington, Martin Throp and Chris Barber (not the jazz trombonist!). Entrants included air-cooled models other than Beetles — Type 3 and Karmann Ghia — which would not be competitive in national championship autotests. Seen here displacing a marker in a VWOC autotest at Newbury is Dennis Pendlebury's 1600TL Fastback.

*Left, below:* Peter Robertson driving his standard 1200 in a VWOC autotest. The stock swing axle rear suspension which caused tricky handling on the road was an advantage when it came to tight turns in autotests.

*Below:* The first VW to make a significant impression on drag racing in the USA was the EMPI 'Inch Pincher', built and driven by Dean Lowry and Darrell Vittone. The car was originally used for slalom and circuit racing and was actually raced by Dan Gurney. Lowry's appearances with the Beetle on drag race strips in 1964 sowed the seeds for a massive industry, producing the most advanced performance components. Classes in US drag racing were based on weight per cubic inch. The Beetle raced in class H, 11lb per cubic inch, competing against much more powerful, but much heavier V8s. 'Inch Pincher' set the records, starting with 14.9 seconds and 91.5mph in 1965. Lowry left EMPI in '68 and formed his own company, coining the best name ever for a tuning firm — Deano Dyno Soars — and building a 2180cc Bug which ran 11.6 seconds and 112mph in 1969. VW engines had raced on the quarter mile before these Beetles; Gene Berg was the trailblazer, building a radically modified engine which powered a single-seat dragster, 'Little But Quick', in 1961. (VW Motoring)

## THE 1960s

*Right:* Speedwell, a name which evokes memories of Graham Hill and racing A35s, became involved in VW tuning in the mid-1960s, bringing into the UK some of the EMPI products from the USA. Speedwell also developed their own twin-carb conversion, seen here, using Stromberg CD carburettors. Also visible are the pipes to an oil cooler in front of the fan. With a 0-60mph time of around 14 seconds, the Speedwell Beetle rivalled Okrasa for performance, at considerably less cost, but one worried about the integrity of the stock VW 1200 crankshaft at high revs.

*Right, below:* The last outright win for a Beetle on the East African Safari Rally was by Tommy Fjastad and Bernard Schmider in 1962, the year when both the Cooper Motor Corporation (the VW importers) and the factory in Wolfsburg put maximum effort into preparation, reconnaissance and servicing to achieve Volkswagen's fourth victory in one of the world's most gruelling motorsport events. Following that triumph, Beetles scored two more class wins in 1964 and 1965. Shown here is the 1965 class-winning Beetle driven by M. Sayeed Khan and Balbir Singh Panue, complete with trophies and battle-scars. (VW Motoring)

*Below:* British rally champion and VW expert Bill Bengry drove one of the Cooper Motor Corporation Beetles on the Safari Rally in 1967, but failed to finish — the only time, he declared, that a VW engine had let him down. The car's failure was due to a severe impact when driving over a washaway, which cracked the gearbox mounting and caused the dynamo and fan to move. The engine had to be taken out and when it was replaced the special dust seal could not be refitted. This led to an accumulation of dust and mud on the cooling fins, causing overheating. (VW Motoring)

# VW BEETLE IN MOTORSPORT

*Left:* Robert McBurney was one of Northern Ireland's top drivers in rallies and autocross. He won the Ulster Rally Championship in 1966 with this Beetle, fitted with a Porsche Super 90 engine, Porsche brakes, stiffer (1948 spec) front torsion bars and Bilsteins. He won autocross championships with the same car. Previously, McBurney had been a class winner with a 1200cc engine on the Circuit of Ireland Rally. (D.B. Crawford)

*Below:* Type 3 VWs were in contention again on the RAC Rally in 1966 and 1967. Sten Lundin and Bengt Karlsson drove privately entered 1600TL Fastbacks to finish first and second in class (eighth and ninth overall) in 1966. The following year Bjorn Waldegaard drove the Fastback shown here to finish 12th overall with absolutely no service, just the wheels and tools carried inside the car!

*Left:* Ken Shields' early motorsport career included a class win on the Circuit of Ireland in 1964, driving a Vauxhall Velox, but he became best-known for his autotest performances in a VW, being virtually unbeatable in the saloon car class. This is Shields driving his first Beetle at a cross-roads autotest in Northern Ireland in 1967. Note the distortion of the rear tyre. (A. McGrath)

50

## THE 1960s

*Right:* Derek Boyd drove a 1500 Beetle on the Circuit of Ireland in 1968, winning the over-1300cc production saloon class (10th overall). Boyd drove the same Beetle for the Northern Ireland team which won the Inter-Area autotests in 1969. *(Brian Foley)*

*Below:* A driver who deserved to win a major championship title in autocross, but sadly never did, was the late Griff Griffiths, a popular, spectacular and innovative VW (and Morgan) tuner/driver. Griff will mostly be remembered competing against Laurie Manifold and Peter Harrold in this 1953 oval-window Beetle with a Type 3 1500S engine and plastic roof. He was one of the first in the UK to do away with the VW's power-absorbing mechanically-driven fan and use an electric cooling fan.

# VW BEETLE IN MOTORSPORT

**Left:** *Griff replaced the VW engine in his autocross Beetle with this Porsche Carrera quad-cam unit. He used the same Beetle in rallycross and later fitted the Carrera engine in a Buggy for autocross. He also drove a VW-Porsche 914 in autocross.*

**Below and bottom:** *The Player's No 6 Autocross Championship final in 1968 was held on the infield area at Silverstone and featured some spectacular jumps. As seen in these shots of Geoff Rosenbloom (108) and Bruce Mankin (112), Beetles always land nose first! Rosenbloom used a Porsche engine and was the class leader in Scotland.*

## THE 1960s

*Above:* This is a genuine split-window Beetle driven in a Chiltern Car Club autocross in 1968 by Ron Johnson. Modifications include a 1600 Porsche engine, additional cooling vents (!), wheel spacers front and rear, and removal of running boards and semaphores (!).

*Right:* Frank Burton was another to add Porsche power to his autocross Beetle, but apparently had some difficulty finding a suitable exhaust system. We're not too sure about the SU carburettors, but it all seemed to work and Burton was a regular class winner in the south-west.

*Right:* Bob Piper's effort tended to provoke boos and hisses from the VW purists. This is a two-litre Ford V4 lump in the back of his Beetle. There's a radiator in the front boot and nasty water-pipes between the two, but despite that it was quite successful in autocross in the south-east. At the time, the only rule about installing different engines was that the number of cylinders had to be the same as the original.

## VW BEETLE IN MOTORSPORT

**Left:** Ken Shields replaced Robert Woodside as the saloon driver in the Northern Ireland autotest team in 1968 and helped the team to eight victories in the Ken Wharton event, invariably scoring class wins himself. From 1968 to 1983, Shields drove for the Northern Ireland team 11 times, always using a Beetle, and was presented with a special commemorative plaque by the Autotest Drivers' Club of Northern Ireland for his achievements. Shields generally ran with a 1600 twin-carb engine, limited slip diff and Konis, and he removed the steering damper. His Beetle is pictured here at the Ken Wharton event in 1969.

**Above and left:** The Sporting VW Club organised trials throughout the decade, with a variety of air-cooled VWs regularly taking part. The Type 3 1500S Karmann Ghia is driven by Ivan Collins and the lhd oval-window '56 sunroof Beetle by Ashley Garbett. Sunroofs were opened to give the passenger more freedom to bounce to assist traction! This event was in 1969. The following year, SVWC ceased to exist as an independent club, becoming the south-east centre of VWOC (GB).

# VWP CAR SALES

**AIR-COOLED VOLKSWAGEN & EARLY PORSCHES BOUGHT, SOLD & PART EXCHANGED
FROM RESTO PROJECTS TO SHOW WINNERS — NO VEHICLE REFUSED**

ANY VW OR PORSCHE CAN BE BUILT TO YOUR SPEC, TO THE HIGHEST STANDARD.

IF YOU REQUIRE A SPECIFIC VW AND IT'S NOT IN STOCK
WE WILL LOCATE IT THROUGH OUR LARGE REGISTERS.

IF YOU REQUIRE A 356 OR 911, PHOTOGRAPHS & VIDEOS
ARE AVAILABLE OF CARS NOT IN STOCK (OR IN OTHER COUNTRIES)
VIEW BY APPOINMENT

## IF YOU WISH TO SELL YOUR VEHICLE

FOR AN OUTRIGHT PURCHASE, OR A MINIMAL COMMISSION FEE, PLEASE CONTACT

### STEPHEN TENORIO on
Office — 0784 469100    Mobile — 0836 204113    Fax 0784 464911
**22 GORING RD, STAINES, MIDDLESEX, TW18 3EH (Office Address)**

VWP vehicles can be viewed at Raven car sales, The Causeway, Staines, Middx
(Resto, Cal and custom Beetles and cabrios)
New branch: 32A Wood End Gardens, Northolt, Middx (Stock Beetles kept at this branch)

---

## BEETLE SPECIALIST WORKSHOP

**Restoration and Renovation to the Highest Standard**

T4 Engine Conversions • Bespoke Engines

Race Prep. • Servicing • Welding

Trials & Off-road prep • Repairs & Resprays

Ballards Place
Eardiston, Tenbury Wells
Worcester WR15 8JR
Tel/Fax: 058 470 348

---

### THE "CARE + REPAIR" CENTRE FOR YOUR BEETLE!

EVERYTHING FROM A TUNE-UP
TO A COMPLETE RESTORATION

**TYPES 1/2 + KARMANN**

(0270) 841971
WYBUNBURY NANTWICH CHESHIRE

---

**INSURANCE APPROVED**

For **GENUINE SERVICE** using only **GENUINE VW**
Audi parts come to the best VW renovation
specialist in the country –

'I put back into VW's what 20 years has taken out'

**Send SAE for FREE CATALOGUE**

TEL: 0932-874848/0932-561659 (Eves)
0836-643578 (Mobile)
**KAR KRAFT**
UNIT 1 SANTA FE, LYNE, SURREY
2 minutes drive from Junction 11 M25

**SPARE PARTS • SERVICING • STRUCTURAL WELDING SPECIALIST**

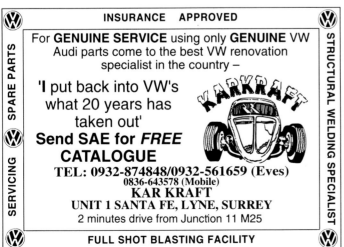

**FULL SHOT BLASTING FACILITY**

---

### L I M I T E D  e d i t i o n

VOLKSWAGEN RESTORATION & PERFORMANCE CENTRE

Suppliers of quality performance and standard parts for the VW Beetle for the past ten years. Over two thousand stock lines, including tuned engines, exhausts, modified suspension components, wheels etc., plus full workshop facilities.

Limited Edition, Warrington Road, High Legh,
Nr. Knutsford, Cheshire WA16 0RT
**Telephone — (0925) 757575**

# The 1970s

## Beetles are champions in European rallycross

In the mid-1970s the Austrian Volkswagen importers, Porsche Salzburg, prepared Beetles for International rallies and rallycross. This was close to being a 'works' team, but had only a very small budget compared with other works teams in motorsport.

The VW factory homologated some motor-sport-orientated parts for the 1302S in 1971, including (in Group 2) a dry sump system with a three-gear oil pump, a five- speed gearbox (from VW-Porsche 914) and an instrument panel with a 0-7000rpm rev- counter in place of the speedometer. A smaller speedo was installed to one side, together with oil pressure and temperature gauges and a clock. A limited slip differential was a standard 'M-series' option (along with items such as laminated windscreen and reversing lights) and so could be used in Group 1.

The Porsche-Austria Beetles ran in Group 2 with 1600cc engines developing about 120bhp. They had Solex 40 PII carbs (later replaced by Weber 46 IDA), with increased compression, big valves and a sports camshaft. As well as the dry sump conversion, five-speed gearbox and limited slip diff, they had front oil coolers and Bilstein suspension.

In 1973 they used both 1302S and 1303S models and entered the Swedish Rally, the Austrian Alpine, the TAP Rally in Portugal and the Acropolis, all events included in the inaugural World Rally Championship. Drivers included Bjorn Waldegaard, Harry Kallstrom and Tony Fall.

Best results were sixth overall by Waldegaard on the Swedish Rally, fifth by Georg Fischer on the Acropolis and tenth place by Tony Fall on the Austrian Alpine.

With insufficient time and money spent on development, there were frequent mechanical failures and the cars lacked the power to be competitive in world-class rallying. But there were one or two isolated successes at this time: in 1974 Werner Koch drove a Beetle to an outright win on the New Caledonia Rally, and in 1976 Leo Schirnhofer was the Group A winner, and eighth overall, on the Acropolis Rally in a 1303S.

The intention was to homologate a two-litre engine, but this was abandoned as a fuel crisis

caused rallying to be halted or curtailed in some countries (including Austria). Porsche-Austria sold most of the rally Beetles, but two 1302S models were converted for rallycross. Type 4 engines (permitted then in rallycross but not in rallying) were installed and were enlarged to 2.4 litres, producing some 180bhp. The cars were extensively lightened and sponsorship was obtained from Memphis International.

After winning several rallycross events in Austria they decided to enter the 1974 European Championship with two drivers, Franz Wurz and Herbert Grunsteidl. The outcome of this decision was to be VW's biggest motorsport success since the Safari Rally: Wurz won the European Championship title, scoring four first places, two seconds, one third and two fourths, beating Waldegaard's Porsche 911 and Blomqvist's works SAAB. Grunsteidl finished seventh in the Championship.

Two other VWs finished in the top ten. Guy Deladriere from Belgium was sixth in a 2.7-litre Porsche-engined Beetle, while tenth place fell to another Austrian, Hubert Katzian, driving a

*Top: Brian Addison and John McKerrell drove a 1302S on the 1971 RAC Rally. Sponsored by the VW dealership Walter Scott, the Beetle was standard apart from Koni shock absorbers, Cibie lamps, sumpguard and roll cage. Amidst all the replacements, rebuilding and rewelding that took place as routine 'servicing' for most rally cars, the Beetle needed nothing more than a wheel change before finishing in 104th place — which was better than 145 other cars that failed to complete the course! The Addison Beetle is seen here in a snowstorm at Castle Howard in Yorkshire.*

*Above: A regular autocross competitor in the '70s was Tony Trimmer, who drove this 1600cc oval-window Beetle, reg. no. HOT 1. Trimmer is seen in action here at a Stort Valley Motor Club autocross in 1971. When Laurie Manifold retired from autocross, in the 1980s, it was Trimmer who bought his Beetle.*

2.4-litre Beetle for the Peter Kiss/Britax team. The top 20 included no less than nine VWs, together with two Porsches, two SAABs, two Ford BDA-powered DAFs, four Ford Escorts and one Opel.

The final round of the 1974 Championship, organised by Thames Estuary Auto Club, took place at Lydden Hill. Wurz won only one of his three races in this round — and in doing so rolled the car after crossing the finish line! The Beetle was not seriously damaged, and by finishing second in the final race behind Waldegaard's Porsche, Wurz scored enough points to clinch the Championship.

Shortly after the 1974 success Porsche-Austria closed down its motorsport department, but the reign of Beetles in rallycross continued, thanks to a Dutch team, Conti Adr van der Ven. The name signifies sponsorship by Continental Tyres and by a garage in Holland owned by Adrian van der Ven. Their cars had three-litre, six-cylinder Porsche engines and the drivers were Cees Teurlings and Dick Riefel.

It was a glorious victory for the Conti Adr Beetles. Teurlings took the 1975 Championship title, with Riefel runner-up. Behind them, though no longer running under the Porsche-Austria banner, Wurz and Grunsteidl drove the same cars as in 1974 (still in Memphis livery) and finished fourth and fifth in the Championship. The leading British driver, Hugh Wheldon, in an Autocavan Beetle with 2.4-litre Type 4 engine, was sixth, and Katzian was seventh. The only car in the top seven that was not a Beetle was John Taylor's Escort, in third place.

In 1976 rallycross rules were modified to outlaw non-original engine blocks, as a result of which neither Porsche nor Type 4 engines could be used. Beetles ceased to be competitive in European rallycross for a few years, though they were to come back again in the 1980s with turbos and four-wheel drive...

Rallycross was not confined to Europe during the '70s; it was also big in Australia. Sponsored by Firestone, rallycross at Catalina Park, in the mountains west of Sydney, was shown on Australian television. A spectacular course featured jumps, watersplashes, and an abundance of mud. Cars ranged from Mini-Coopers to five-litre Holdens, but the winners more often than not were Beetles. Using 2.2-litre engines, VW drivers Chris Heyer, Peter Mill, Barry Ferguson and Doug Chivas were regular winners at Catalina, filling the top places in the Australian Rallycross Championship. Ferguson's Beetle won more Division 1 rallycrosses than any other car, gaining exciting victories against the supercharged Holden Torana of Larry Perkins (son of the '50s Round Australia Rally winner, who went on to become a successful driver-engineer-team leader in touring car racing) and Peter Brock (who is one of the famous names of Bathurst, having won The Great Race nine times).

In British autocross, Laurie Manifold continued to be one of the great names, but came increasingly under challenge from Peter Harrold, a Norfolk farmer, and his brother, Paul. The Harrolds worked closely with Autocavan, the country's leading air-cooled VW tuning firm. Autocavan was the first to have two litres and 48 IDA Webers running in motorsport in the UK, and Paul Harrold became champion of the rear-engined class in autocross in 1971, using an Autocavan engine.

Manifold, who was then only up to 1884cc, rolled his Beetle spectacularly in the BTRDA final, in an incident that was not due to driver error. In an effort to match Harrold's greater speed on the long straights, Manifold had fitted 15-inch wheels instead of his usual 13-inch. These had narrower rims and, when cor-

## THE 1970s

nering, picked up stones between tyre and rim which caused the tyre to deflate and consequently the roll.

The Harrold brothers had their share of drama the following year when their Beetle caught fire on its trailer behind the tow car, only three weeks before the BTRDA final. They built a new car just in time, and Peter won his championship class.

Manifold received an offer of a super-powerful engine built in Finland for ice-racing. To circumnavigate import restrictions of the time, Laurie bought a new VW, installed a clapped-out old engine, and took it to Finland along with his mechanic, Roy Wilson. There they swapped the old engine — plus several pockets full of fivers — for the racing engine. They disguised the Beetle as a rally car with mud and stickers and shipped it back to England. Its first outing was on a scorching hot August Bank Holiday during which the engine, designed for sub-zero temperatures, almost literally melted. 'After that,' Manifold recalls, 'we decided to get our engine components from America.'

When Manifold switched from a 1500 Beetle to a 1302S, his old car was bought by Brian Prior. A 'match' was arranged between the two cars at a Stort Valley autocross, and when the flag dropped Manifold in the new highly-modified 1302S was beaten off the startline by Prior in the old swing-axle Beetle. Prior pulled out a lead of two seconds per lap, causing much consternation in the Manifold camp and ribaldry from Prior's supporters. Manifold's explanation (excuse?) was that there had not been enough time to install his usual fly-off handbrake in the new car, and as he forgot to release the ratchet he had run the entire race with the handbrake on!

Another incident occurred when Manifold had to replace the exhaust between races and left routine chores such as checking petrol level to someone else, who apparently left the filler cap loose. Laurie remembers the next race: 'As I yumped round the particularly bumpy circuit flames crept up my leg and I could see in my mirror a trail of blazing grass. I jumped out damn quick.' The cap had fallen off, petrol had splashed out and been ignited by sparks from the battery (always transferred to the front to improve weight distribution).

Peter Noad ran a modified 1500 Beetle, reg. no. SKT 328H, in rallies and autotests, sponsored by *Cars and Car Conversions* magazine. Outright rally wins included Birmingham University Motor Club's Welsh Mermaid, Coventry and Warwickshire Motor Club's Precinct Rally, the Telehoist Motor Club Kilwin Rally and the Dursley MC Hi-Fi Rally. As rallied in 1971, the engine was 1678cc with a single Minnow-Fish carb. Subsequently it was fitted with twin Solex 40 PII carbs, on Sauer manifolds, after which further developments by Autocavan included an enlargement to 2074cc with a roller bearing crank.

Special mods introduced by Noad for rallying included re-locating the screenwasher reservoir. The standard VW screenwasher was a pneumatic device, which worked very well but was attached to the spare wheel. In the event of a puncture, the driver would throw the spare wheel out of the boot, which pulled the screenwash reservoir off its pipe and released several pints of water at 40psi into the driver's face!

Other features included the towbar (which acted as a sumpguard), an electric dipstick (oil temperature gauge) and a modification to the accelerator pedal. The standard Wolfsburg design required that the driver's leg be three inches shorter when operating the brake than when using the accelerator.

Noad's major successes in this Beetle were

in autotests, including victory in the RAC National Championship in 1972. In the same year SKT 328H also picked up the BTRDA Westwood Trophy and three regional autotest championships — Central Southern, London Counties, and Midlands, in effect the 'grand slam' of autotests. Participation in 84 events brought 55 class wins, of which nine were outright fastest times. Main opposition in the saloon class came from MG 1300s, Escort Mexicos and RS 2000s.

Dennis Greenslade first drove a Beetle on a classic trial in 1971. He won that event, the Maidstone & Mid-Kent Motor Club Tyrwhitt-Drake Trophy, outright. Since then, and up to 1992, Greenslade has won 229 awards in 205 trials, the reason for the awards outnumbering the events being that he wins team awards and championships. He won MCC Triple Awards in 1971, 1973 and 1979, and won the VWOC Classic Trials Championship every year from 1974 to 1979.

During the 1970s Greenslade took part in more than 150 trials in his 1302S, reg. no. 2373 VW, with one hundred percent reliability. He failed to finish only twice — once when he collided with a tree on Rocombe Hill on the Exeter Trial, and once when he suffered two punctures (with only one spare) on the Exmoor Clouds Trial. Greenslade recalls how he persuaded a passing German tourist in a Beetle to lend him a wheel. The German was not even carrying a spare but agreed to wait, with his car jacked up on three wheels, while Greenslade took off to do the next trials section on the borrowed wheel. In the end it was all in vain: they had fallen too far behind schedule and the marshals had closed the section.

In 1978 Greenslade acquired a Brasilia which he used for a time in classic trials, winning the David Paull Trophy which was effectively a National classic trials championship. He also owned a second Brasilia in which he set a new record in 1978 for the Land's End to John o'Groats return journey, completing it in 32 hours and 44 minutes.

Ken Shields drove a Beetle as a member of the Northern Ireland team in the Ken Wharton Memorial Autotests for most of the 1970s. Northern Ireland was the winning team, and Shields the saloon class winner, in 1969, '70, '71, '72, '73, '75 and '76. His fellow team members were Harold Hagan (MG Midget) and Ken Irwin or John Lyons (Mini-Cooper S). During those years Shields won the large saloon class at every Northern Ireland autotest (plus some in England and Scotland) and he was the autotest saloon class champion of Northern Ireland for eight years.

Shields also drove his Beetle in autotests in Sweden, where he had his only mechanical problem with the car. The event took place in a multi-storey car park, with test manoeuvres on every floor. Continual high-speed clockwise turns up the spiral ramp caused all the oil to migrate to the rocker box on one side of the engine, causing it to seize when Shields reached the top level.

Home-made VW-based specials and buggies were widely used in autocross, trials and midget car racing in Ireland. David Sheane, who had been building very successful lightweight VW trials specials for many years, decided to build a Formula Vee car which he used in hillclimbs. In 1975 he was instrumental in starting Irish Formula Vee and is now one of the principal constructors of Vees raced in both Ireland and England.

In Germany a notable motorsport Beetle was the 1302S of Dieter Götting who raced with considerable success in slaloms and hillclimbs. The Beetle was a 1600cc class winner. Considered very radical in the early '70s, Götting's engine used ram air cooling (no fan), had the dry sump system (with 11 litres oil capacity), Porsche 914 five-speed gearbox, twin Weber 48 IDA carbs and 11 to 1 compression. It produced 158bhp at 7600rpm! Quoted performance was 0-100kph (62.1mph) in 6.5 seconds.

In the USA Beetles started dominating their classes in drag racing. In 1971 Darrell Vittone built a successor to the original EMPI Inch Pincher. Known as Inch Pincher Too, this Beetle had its roof lowered by nearly four inches and a two-litre engine using state-of-the-art EMPI components to develop about 170bhp. It became the quickest I/gas class car in the country, running 11.76 seconds and 113mph in 1972.

US drag racing classes were changed in the 1970s, making it harder for VWs to race against domestic front-engined sedans. Attention swung towards all-VW events, known as Bug-Ins. More radical cars appeared; one of the first to do away with the VW floorpan and use a tubular space-frame was Bill Mitchell, who ran under 11 seconds in the mid-'70s.

The magazine *Dune Buggies and Hot VWs*, founded in California in 1967, introduced comprehensive coverage of both drag racing and off-road racing and, to the present day, remains 'the Bible' for these two specialised branches of the sport.

The Midwestern Council of Sports Car Clubs (which covers the Wisconsin and Illinois areas of the USA) introduced a VW class in club racing in the mid-'70s, but with only about six active competitors it was short-lived. Art Schmidt was racing a Beetle at that time and later found himself in various different classes with other makes as the rules kept changing. As well as road races at such circuits as Wilmot and Lynndale Farm, the sports car clubs staged hillclimbs, autotests (known as gymkhanas in the USA) and slalom-sprints (which the Americans call autocross).

VWs did not play a significant role in British drag racing during the 1970s. There was one 'trailblazer', though: Bernie Smith, proprietor of Wagenmaster, in Wisbech, was a runner-up in the 1974 British Drag Racing Association Championship, with a Beetle known as Humbug. Times were then in the 16-second bracket.

VW tuning really took off in the 1970s. Most of the components originated in the USA from firms such as Gene Berg, Deano Dyno Soars, Scat and Bugpack, and were aimed primarily at drag and off-road racing. There were already long-stroke crankshafts and big bore kits to enlarge capacity to more than two litres, big twin-choke Webers, big valves and hot cams. But there were now also such innovations as high-ratio rockers, sump extensions, small pulleys to reduce fan speed (the fan loses efficiency above the stock rpm range), straps to support the gearbox and prevent the front mounting breaking (which it tended to do, making the car rather difficult to drive with the engine only held in place by the clutch cable), and much more. After-market cylinder heads and crankcases, for racing, also began to appear — leading towards the present situation where a full race 'VW' engine can be built using no original Volkswagen components whatsoever.

During this decade the Volkswagen factory came up with McPherson strut front suspension (1302S), twin port cylinder heads (previously only Type 3), rack and pinion steering and a curved windscreen (1303). In July 1974, production of Beetles ceased at Wolfsburg. Production continued in Emden, Germany, until January 1978, and thereafter in other countries, notably Brazil and Mexico.

**NOTTINGHAM'S No.1 BEETLE WORKSHOP**
Members of Motor Vehicle Repairers Association

Restoration
Convertibles
Cal look
Sales
Service

**Watnall Road, Hucknall Nottingham**
 **0602 681504**

5 mins from Junction 26/27 off M1 Motorway

---

# HOUSE OF HASELOCK 'THE COUNTRY'S LEADING VW RESTORATION SPECIALIST'

### HOW RESTORATIONS CAN SAVE YOU MONEY

There's nothing to touch your VW in terms of quality but the British climate corrodes any car's bodywork eventually. You could replace your beloved VW with a new car – but new car quality can be a joke! What's more, a Haselock restoration preserves the value of your Beetle or Caravette (or other Type 2) whereas a newer car loses money the moment you drive it away. So let us build back into your car the quality put there by VW's engineers in the first place. We've been restoring VWs for 20 years and between us, we've spent 65 years working on them! It makes you feel old!

THIS – TO THIS . . .

ALL BEETLE & TYPE 2 PANELS SUPPLIED & FITTED, IN-CLUDING SOME MADE UNIQUELY FOR OURSELVES

**COLLECTION & DELIVERY SERVICE – NATIONWIDE!**

☆ FULL OR PLANNED STEP-BY-STEP RES-TORATIONS AVAILABLE

☆ QUALITY *MOT WELDING* ON ALL MODELS, INCLUDING GOLF

☆ SPECIALIST SUPPLIERS AND FITTERS OF ALL AIR-COOLED AND WATER-COOLED ENGINES

☆ ADVICE, INSPECTION AND CONSULTA-TION FREELY GIVEN – PLEASE SEND SAE WITH ALL WRITTEN ENQUIRIES

**HASELOCK**

22 Slingsby Close, Attleborough Fields Ind. Est.,
Nuneaton, Warwickshire, CV11 6RP.
NUNEATON (0203) 328343
Fax: (0203) 385218
**OPEN DAILY 9.00–5.00, SAT. MORNING,**

---

# INJECTION PROBLEMS?
### No Problem For
# Machtech

**V.W. AUDI & PORSCHE SPECIALISTS SINCE 1973**
*ROLLING ROAD DIAGNOSTIC TUNING*
**ULTRA-SONIC INJECTOR CLEANING – FULL SERVICING & ALL REPAIRS**

Unit 2, Happy Valley Industrial Park
Primrose Hill, Kings Langley
Herts. WD4 8HD  (Junction 20, M25)

**KINGS LANGLEY (0923) 269788**   MON-FRI 9am-6pm

---

THE CARRIAGE WORKS,
HEATH END ROAD,
NUNEATON CV10 7JB
(0203) 350766

ALL PARTS AVAILABLE, TOO MANY TO LIST, PHONE WITH REQUIREMENTS
WORKSHOP FACILITIES
FROM A WING TO A FULL RESTORATION NO TASK TOO SMALL OR LARGE
IF YOU HAVE AN IDEA IN YOUR HEAD WE ARE SURE WE CAN TURN IT INTO A VW TO BE PROUD OF. OUR VEHICLES HAVE WON TROPHIES AT ALL MAJOR EVENTS IN 1991 AND 1992
PLEASE CHECK AVAILABILITY BEFORE TRAVELLNG
SO GIVE US A CALL OR SPEAK TO US AT THE SHOWS THIS YEAR,
PHONE SIMON ON (0203) 350766
ALL PRICES INC VAT.

PHONE FOR CARRIAGE CHARGES

---

**CARPET SETS   DOOR PANELS   SEAT COVERS**

FOR BEETLES AND CAMPERS

PHONE FOR DETAILS

**D & M MIDDLETON & SON**
Rawfolds Mills, Cartwright Street, Cleckheaton,
W. Yorks BD19 5LY
Telephone: 0274 871509   Fax: 0274 869950

---

# MICRO GIANT
**VOLKSWAGEN & PORSCHE ENGINEERS**

## 30 YEARS EXPERIENCE!!

From a Kubelwagon to a Corrado.
No matter what your vehicle, let us give it the ultimate care and attention.
*All servicing, repairs, modification & restoration.*
*Difficult jobs our speciality.*
Full range of new & S/H spares. Standard & Modified.
Full range of machining services.
Engine and transmission building, any age, any spec.
**From a shopper to a screamer.**
**VW Drag Racing Champions 1990 & 1991**

Unit 7, Westfield Close, Rawreth Industrial Estate, Rawreth Lane, Rayleigh, Essex
Tel: 0268 782601

## THE 1970s

*Right:* Lee Lucas drove a GP Buggy as a member of the Northern Ireland autotest team. Pictured competing in the Inter-Area team event in 1973, Lucas' Buggy had a twin-carb 1600cc engine, limited slip diff and 13-inch wheels. Compared to purpose-built autotest specials, the VW Buggy is rather bulky. As specials had a five percent handicap applied to their times, the Buggy's success depended on how it was classified: as special or sports car. On scratch times, Lucas was certainly capable of outright wins.

*Right:* Steve Stringer was a class winner in the South East Regional Autotest Championship in 1973, driving this 1303S. Modifications included twin Minnow-Fish carbs. Stringer subsequently became national champion, driving a Lotus 7.

*Below:* Although VW drag racing did not get serious in the UK until the first 'Bug Jams' in the late '80s, a few quarter-mile enthusiasts raced VWs long before that. These two road-going Beetles were pictured at Santa Pod in 1973. The 1302S achieved 19.7 seconds. (VW Motoring)

# VW BEETLE IN MOTORSPORT

*Left: From 1970 to 1978 Peter Keegan's VW won practically every drag race entered in Australia and held five national records, achieving 10.29 seconds and 134mph. Keegan's Beetle was the first four-cylinder sedan in the world to go under 11 seconds for the quarter-mile. The 2180cc engine was fitted with a Godfrey Marshall supercharger which was actually an aircraft cabin air-conditioning blower. The car was built with all Australian-made parts, except for the SPG roller crank and original VW components. (P. Keegan)*

*Below: SKT 328H was the 1970 model 1500 Beetle in which Peter Noad won ten autotest championships, including the RAC national title. The engine was modified initially by Cartune and later by Autocavan. As competing here in the Felixstowe round of the Castrol-BTRDA Championship in 1974, the Autocavan engine was 2.1 litres with Solex 40 PII carbs.*

*Left: Peter Noad giving his impression of jazzman Roland Kirk. The three different types of tailpipes for the Beetle silencer had significant effects on noise and performance. Noise was on the limit for road rallies in the 1970s and Noad kept the standard 'pea-shooter' tailpipes in the boot, to be used if the Beetle failed the noise test at scrutineering with the Taper Tips fitted. The latter improved 0-60mph time by one second but increased noise by 4dB.*

THE 1970s

*Above:* Standard apart from a Minnow-Fish carburettor and Konis, TKE 367H was Peter Noad's 'other' 1500 Beetle, deputising when the 2.1-litre was out of action and contributing to championship wins in 1973 and 1974. TKE had previously won the BTRDA Westwood Trophy and the Central Southern Autotest Championship in 1970, altogether notching up 40 class wins and six overall fastest times.

*Right:* Minnow-Fish carburettor on an otherwise standard 1600 Beetle engine increased power by 6bhp on a rolling road and improved 0-60mph acceleration by two seconds. Minnows were used by several successful VW motorsport drivers in Britain.

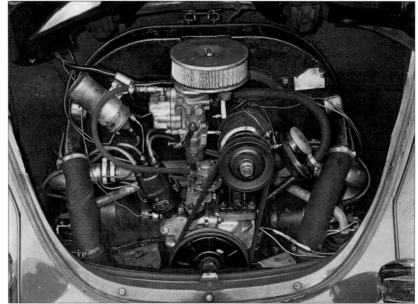

*Right:* In 1972 Paul Raymond ran one of the first turbocharged Beetles in rallying. Designed by Bob Henderson, the Minnow genius, the turbo blew through a Minnow-Fish carburettor. Apart from the Minnow carb and turbo the engine was standard, producing 120bhp and propelling the Beetle from 0 to 60 in just over nine seconds. Raymond scored an outright win on the Ebworth Chase special stage rally, but sadly was to lose his life in an air crash.

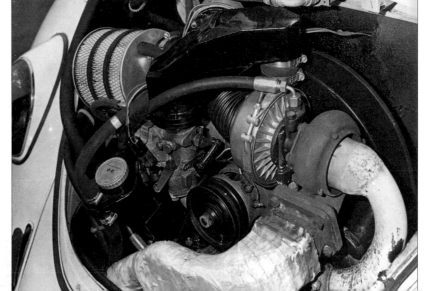

VW BEETLE IN MOTORSPORT

# THE 1970s

*Opposite page*

**Top:** *The Volkswagen factory in Brazil produced this sports coupe, known as SP2, in the mid-70s. Based on a Karmann Ghia floorpan, it was powered by a twin-carb Type 3 engine, enlarged to 1700cc. There are no records of the SP2 being used in motorsport outside its own country, but some of the components produced by VW in Brazil exclusively for this model (eg cylinder heads) are now used in preference to original German VW parts for tuned Beetles in motorsport.*

**Bottom:** *The Austrian VW importers ran a team of Beetles in International rallies in 1973. Drivers included Harry Kallstrom, seen here in action on the TAP Rally in Portugal. His was one of four VWs on the rally, all of which retired. Kallstrom had throttle linkage failure, Georg Fischer gearbox trouble, Tony Fall had a back wheel (and drive shaft) come off, and Herbert Grunsteidl crashed!*
*(Colin Taylor Productions)*

**Top:** *Tony Fall drove a Porsche-Austria 1303S to tenth place on the Austrian Alpine Rally (a round of the World Championship) in 1973. Fall, a successful works driver with British Leyland and Lancia, pronounced the Beetle great fun to drive: 'you could get away with incredible manoeuvres that were not possible in other cars, and just let it career between the trees. But there was not enough power to change the angle of attack (in the middle of a corner) and not enough development'.*
*(Colin Taylor Productions)*

**Middle:** *Swedes Carl-Eric Larsson and Fergus Sager drove this 1302S on the RAC Rally in 1973. Seen here on the first day in Clipston Forest, they retired after hitting a bank in Yorkshire on the last day.*

**Right:** *One of several marathon rallies staged at irregular intervals was the World Cup (sponsored by UDT) in 1974. It started in Wembley and finished in Munich, covering 11,000 miles in between, crossing the Sahara twice and visiting Nigeria. Among 52 starters there was just one air-cooled VW, a Brasilia crewed by Carlos Weck and Claudio Mueller, which finished in 16th place. The Brasilia is a hatchback made up of various Beetle, KG, Type 3 and local Brazilian-made components, powered by a 1600 Beetle engine.*

# VW BEETLE IN MOTORSPORT

**Above:** In 1974 Porsche-Austria converted two of their 1302S rally Beetles into rallycross cars and entered the European Rallycross Championship. The cars were considerably lightened and fitted with Type 4 engines.

**Left:** Rallycross Beetle Type 4 engine is 2.4 litre with twin Weber 46 IDA carbs. Cooling is by means of a Passat electric fan. Power output was quoted as 180bhp.

## VW BEETLE IN MOTORSPORT

*Right:* Ron McFarlane autocrossing his roadgoing 1500 Beetle at a Chiltern Car Club event in 1968.

*Below:* Northern Ireland team member Derek Boyd in a 1500 Beetle at the 1969 Inter-Area autotests.

*Bottom:* B. Sand in a 1200 Beetle climbing a sandy section at Canada Heights in a Sporting VW Club trial, 1970.

VW BEETLE IN MOTORSPORT

*Above:* Autocross maestro Laurie Manifold in action in 1971. He rolled this car in the BTRDA Championship final, and it was subsequently adorned with a 'This Way Up' sticker.

*Left:* 1975 European Rallycross champion Cees Teurlings in action at Lydden Hill in his six-cylinder Porsche-engined Beetle.

*Below:* The fastest street-legal Beetle in Britain in the 1980s was John Brewster's 2.1-litre Autocavan 1303. Brewster's record for the quarter-mile was 12.27 seconds in 1989. (M. Key)

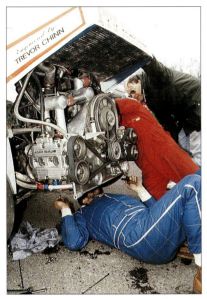

*Above:* Rallycross produces the ultimate in motorsport Beetles and they do not come any more ultimate than Peter Harrold's Autocavan/PPJ Team 16-valve, turbocharged, four-wheel-drive flyer.

*Above, right:* Perry Andersson's 1500S Notchback was very impressive on the 1992 Charrington's RAC Historic Rally.

*Right:* Bob Beales and crew servicing his historic rally Beetle. Beales won the Welsh Historic Championship in 1991 with his '58 Okrasa Beetle.

*Below:* Phil Horne and Simon Howarth in a Big Boys' Toys Beetle Cup race at Cadwell Park in 1992.

VW BEETLE IN MOTORSPORT

**Above:** David Lucas (navigated by his father) competing on the 1982 Lombard RAC Golden Fifty Historic Rally in a 1957 1200 Beetle. (Foster & Skeffington)

**Left:** Multiple trials champion Dennis Greenslade churning through the mud on the Clee Hills Classic Trial in his 1300 Beetle.

# VW BEETLE IN MOTORSPORT

*Above: Getting out of a trials section can be difficult. This is Geoff Margetts needing a few helping hands from VW Owners Club marshals.*

*Right: The 1991 BTRDA Production Car Trials Champion, Mike Stephens, negotiates an adverse camber at a Sporting Owner Drivers Club trial at Great Brickhill, in his 1302S.*

*Below: Arthur Vowden (and his son Dean) designed and built this Trials Special, using a shortened Beetle chassis and Type 3 engine. Vowden Sr. won the ACTC Classic Trials Championship in 1991.*

VW BEETLE IN MOTORSPORT

*Above: A 1968 Beetle with 2.1-litre engine, built by Vintage Vee-Dub Supplies for the 1993 London-Sydney Marathon, was driven by George Geshos and Boris Orazem on the Targa Tasmanian Rally, coming home second in class. (Vintage Vee-Dub Supplies)*

*Left: Käfer Cup competitor Ralf Schaub driving his very nice oval-window Beetle on the hillclimb at Trier, Germany, in 1991. (Gute Fahrt/Joachim Fischer)*

*Below: Walter Schäfer, winner of the Käfer Cup in 1989 and 1991, in action in a slalom at Nordlingen, Germany. (Gute Fahrt/Joachim Fischer)*

VW BEETLE IN MOTORSPORT

*Above:* Richard Hölzl has finished in the top three in sports sedan racing in Australia (competing against other makes including V8s) in his road-registered 2.3-litre oval-window Beetle. (G. Aungle)

*Right:* There will always be Beetles painted to look like the Walt Disney 'Herbie'. This number 53 is being raced by Käfer Cup competitor Klaus Baumann in a slalom at Nordlingen in 1990. (Gute Fahrt/Joachim Fischer)

*Below, right:* Dave Perkins' Pro Turbo Beetle at L.A. County Raceway in November 1991. Perkins was the first to achieve 150mph in a Beetle in drag racing. (K. Seume)

*Next page:*
Drag race driver Paul Hughes likes to be known as 'The Poser', but it is certainly no pose to be sitting behind the 682bhp Autocraft engine inside Brian Burrows' 'Outrageous' Beetle Funny Car at 160mph.

VW BEETLE IN MOTORSPORT

## THE 1970s

*Right:* The Austrian rallycross Beetles were driven by Herbert Grunsteidl, pictured here at Lydden Hill in 1974, and Franz Wurz. It was one of Volkswagen's biggest motor-sport successes, with Wurz winning the European Championship title. Grunsteidl came seventh.

*Below:* Wurz in action in 1975. Porsche-Austria closed their motorsport department at the end of 1974, but Wurz and Grunsteidl kept the cars, retaining Memphis sponsorship and enlarging the engines to 2.7 litres. In the 1975 Championship, they took fourth and fifth places.

*Right:* Beetles won the European Rallycross Championship again in 1975. This time it was a Dutch team, entered by Adrian van der Ven, sponsored by Continental Tyres and powered by Porsche. The champion driver was Cees Teurlings, pictured setting fastest time of day at Lydden. Runner-up was team-mate Dick Riefel in a similar Beetle.

## VW BEETLE IN MOTORSPORT

*Top:* The Conti Adr. van der Ven 1303S Beetles had three-litre six-cylinder Porsche engines, prepared by Kremer, and giving about 280bhp. Other features included a Porsche gearbox (only four-speed) with limited slip differential and Bilstein suspension. 0-60mph acceleration was quoted as 3.2 seconds.

*Below:* Woud Couwenberg was another Dutch Beetle driver in the top ten of the European Rallycross Championship. His VW was powered by a two-litre Ford BDA engine installed in front of the rear axle. Note the air intakes.

*Bottom:* Interior of Guy Deladriere's Porsche-engined rallycross Beetle, showing an unusual oil cooler installation, inside the cockpit, with an electric fan fitted in the door.

**Opposite page**

*Top:* Leading British entry in the 1975 Rallycross Championship was Hugh Wheldon in this Autocavan-prepared 1303S. The engine was a 2.4-litre Type 4 which ran with ram air cooling (no fan) via ducted intakes through the rear side windows. Wheldon gained sixth place in the Championship.

*Bottom:* In 1976 Beetles were not allowed to use Type 4 or Porsche engines in rallycross. Most of the top VW drivers switched to Porsche 911s (or in Wurz's case a Lancia Stratos) but John Button persevered with a Type 1-engined Beetle, with help from Autocavan. Button had won the National Autocross Championship class in 1974. He moved on to rallycross, the engine was enlarged from 2.1 to 2.3 litres, and he won the Embassy British Rallycross Championship in 1976.

## THE 1970s

## VW BEETLE IN MOTORSPORT

**Above:** Dennis Greenslade is probably the most successful VW driver in classic trials. He has gained a record six MCC Triple Awards and won the VWOC Classic Trials Championship eight times. In the early 70s Greenslade's 1302S Beetle had few modifications apart from a Nikki carburettor, Brabham exhaust and suspension modified to give more ground clearance. In 1974 the engine was enlarged by Cartune to 1776cc, the Nikki replaced by a Reece-Fish, and Konis were fitted.

**Left:** Dennis Greenslade (right) in rather different attire from the trials driver's usual muddy Barbour suit and wellies, receiving the David Paull Trophy, presented by the Duke of Westminster, for his victory in the Classic Trials Championship in 1979. The Duke was patron of Chester Motor Club, the Championship organisers. David Paull, who built engines for trials cars, was a sponsor. (D. Greenslade)

## THE 1970s

*Above:* Paul Collard getting his 1600 Fastback to the top of a steep section in a VW Owners Club trial at Houghton Conquest, Bedfordshire. Type 3 VWs can be very competitive in classic trials provided there are no tight turns where this model's large size and poor turning circle would be a handicap.

*Right, above and below:* From 1976 to 1982 Brian Betteridge was very successful with a VW in autocross. He scored 76 class wins, many of which were outright FTDs, won the BTRDA Championship twice, and numerous regional championships in the Midlands and North West. Initially he ran with a 2.1-litre Type 1 engine, built by Phil Marks Engine Developments. A Brazilian gearbox was used, with Hewland internals, specially-designed drive shafts and Koni suspension. Later Betteridge's Beetle was fitted with a Ford BDA engine.
(B. Betteridge)

# VW BEETLE IN MOTORSPORT

*Above:* In 1977 the London to Sydney Marathon, sponsored by Singapore Airlines, became the longest-ever rally. It covered 20,000 miles, travelling through Western Europe, Iran, Afghanistan, India and Singapore, and culminating in 7,000 miles round Australia. Two Beetles were among the 77 competing vehicles which inclued Porsches, Jeeps and even a five-ton Leyland truck! Francis Tuthill and co-driver Anthony Showell took a then five-year-old 1302S, together with a vast load of spares including a complete spare 1800cc engine — which they needed before leaving Europe! Numerous incidents and repairs left them with time penalties equivalent to several days, but they qualified as finishers in 36th place. Tuthill drove the same Beetle on the 1979 RAC Rally and finished 56th.

*Left:* The Franco-German crew of J. Jeandot and W. Koch took this 1200S Beetle (1600- engined 1200) on the London-Sydney event. They had no spare engine (less weight and more faith!) and finished a highly creditable 17th.

## THE 1970s

*Right:* David Lucas competed in all the home International rallies (including the Welsh, the Manx and the RAC) in this almost-standard 1302S. Here he is seen on the last stage of the RAC, at Wykeham Forest, in 1974. Co-driver was Dennis Abbott.

*Below:* On the Mintex Dales International Rally in 1974 Lucas and Abbott finished 2nd in class (Group 1 up to 1600cc) in their 1302S. (D. Lucas)

# VW BEETLE IN MOTORSPORT

*Left:* Competing here on the 1976 RAC Rally is the privately-entered 1300S (GT Beetle) of Ernst Hubertus. He was a regular finisher, this Beetle completing several RAC Rallies. It was competitors such as Hubertus, Lucas, Tuthill and Addison who kept the VW flag flying on the RAC during the period between Scania Vabis in the 1960s and the Golf GTIs and Audi quattros in the 1980s.

*Below:* After winning the National Autocross Championship in the 1970s, Peter Harrold progressed to rallycross, steadily developing his Autocavan Beetle and adding a turbo and 4wd in the 1980s. At the BTRDA Rallycross Championship round at Snetterton in 1979 (pictured), Harrold's Beetle was still as it had been for autocross, with normally-aspirated 2.1-litre engine. In very muddy and slippery conditions, Harrold was 11 seconds faster than anyone else on one run, and finished as outright winner.

# Stateside Tuning

*Racing and Rallying to International since 1977*

*All Work undertaken incl Preparation for Motor Sport*

- Sole agents for Eurorace, SCAT, G-Berg, BAS
- Full range of Performance Parts and Accessories
- Type I 30HP incl OKRASA to Full Race Spec 3.0 Ltr, Type IV
- Full range of Dellorto Carb Kits and all Replacement Parts and Jets
- **Full Workshop and Machining Facilities**
- Special Case and Head Work • Balancing and Lightening • Aluminium Welding • Valve Guides • Seats • Align Bore (While U Wait) we will machine any VW and Porsche engine
- Suspension
- Full range of Koni and KYB Sports Shox, H/D Anti-roll Bars, Special VAG H/D Rubber and Urethane Mounts
- Wheel Alignment • Ball Joints and Race Suspension Set-ups at Competitive Rates
- Beetles & Porsches bought & sold
- Porsche Fan Conversions for Type I and Type IV Flat Carrera Style also available
- Dry Sump Conversions

**Stateside Tuning,**
*VW, Porsche Audi Specialists*
Unit 3, Enterprise Works,
Alexandra Road,
Enfield, Middx. EN3 7EH
Tel: 081-805 4865/Mobile 0860 888535

---

## BOLAND ADVERTISING

### THE EFFECTIVE SPACE SELLING AGENCY

WE SELL SPACE ON SOME OF THE COUNTRY'S LEADING MAGAZINES AND WE BELIEVE IN A WARM AND FRIENDLY SERVICE, SUCCESSFULLY MIXING CARING DEDICATION AND EFFICIENCY

CONTACT COLETTE BOLAND · BARRY WADSWORTH-SMITH
**BOLAND ADVERTISING & PUBLISHING**
SALATIN HOUSE, 19 CEDAR HOUSE, SUTTON, SURREY, SM2 5JG
TEL: 081 770 9444 FAX: 081 770 0102

---

# VW BEETLE & CABRIOLET AGREED VALUE INSURANCE

## ARRANGED AT LLOYDS

**Lowest Rates for Comprehensive (inc. Karmann Ghia)**

Evidence of No Claims Bonus NOT Required

Premium and availability of cover depend upon individual assessment

Details from:
**N & S SOUTTER & CO.**
'Freepost', Maidstone, ME14 1BR
TEL: Maidstone (0622) 690650

# The 1980s

## *Back to the roots; classic trials and historic rallies*

The only form of motorsport which has hardly changed over the decades, and indeed has no need to change, is trialling. Some of the classic sections such as Simms Hill (near Ilsington on the edge of Dartmoor), Meerhay (at Beaminster in Dorset), Crooked Mustard (near Dursley in the southern Cotswolds) and Beggar's Roost (near Lynton, North Devon) were first used as long ago as the 1930s. Another famous classic trials hill still in use is Bluehills Mine, on the MCC Land's End Trial, which follows an old miners' footpath to the top of the cliffs overlooking the Cornish town of St Agnes. A section on the VW Owners Club Clee Hills Trial, known as Jenny-Wind, goes straight up a 45-degree hillside on the site of an old cable-car lift at Much Wenlock, Shropshire. Nailsworth Ladder is a 'stepped' climb which shortcuts a loop of tarmac road out of Nailsworth village and gives an almost aerial view of the rooftops. It is used on the Cotswold Clouds Trial.

Some of the cars which compete on trials are also 50 or 60 years old. MG J2, Singer Le Mans, Ford Model A and Austin Chummy are names which still appear in the entry lists of MCC trials today.

In trialling terms, the VW Beetle is a modern car. It is the *only* modern production car able to compete alongside purpose-built trials 'specials' such as Dellows, Cannons and Trolls. Of course, Austin-Healey Sprites, Ford Escorts, Vauxhall Novas and others can win awards in classic trials, but the sections are made less difficult for these classes of cars by moving the start-line back, omitting stop-restart lines, and so on. Beetles, especially those with engines over 1300cc, are often required to stop and restart halfway up hills to make the tests more demanding.

The number of Beetles in classic trials has steadily increased, and by the 1980s VW had become the most popular make. Typically, MCC trials have 200 car entries, of which at least 30 are Beetles. Also to be found are Type 3 VWs and VW-engined specials and Buggies.

*Right: Beetles made a comeback in rallycross in the 1980s. The first of the Turbo Beetles to be really competitive at top International level was that of Lars Nystrom, from Sweden, who took eighth place overall in the Motaquip Rallycross Grand Prix at Brands Hatch in 1982. In fact Nystrom was initially well in contention for second place, battling against Matti Alamaki's Porsche and Martin Schanche's awesome Ford Escort Turbo. Stig Blomqvist drove this Beetle in a rallycross event in Sweden.*

# CALLING ALL VW OWNERS!

## The Association of British VW Owners Clubs invite you to join

## BRITAIN'S LARGEST VW ORGANISATION

\* SOCIALS \* NATIONWIDE DISCOUNTS \* INSURANCE \* MEETS \*
\* CAMPING \* CHEAPER RAC MEMBERSHIP \* ROAD EVENTS

Let the ABVWC cut YOUR VW motoring costs!!

**Apply to: Dept V**  66 Pinewood Green, Iver Heath, Bucks SL0 0QH
Tel: 0753 651538

Statistics show the VWs to be more reliable and more successful than other makes. To quote a typical example: on the Exeter Trial in 1987, 21 percent of the cars entered failed to finish, but only 12 percent of the VWs retired. And, while the overall percentage gaining first class awards was nine percent, for the VWs it was 12 percent.

The Association of Classic Trials Clubs was formed in 1979, reflecting the increasing popularity of 'back to the roots' motorsport, and with the aim of preserving the 'amateur' spirit of the sport. In 1984 the ACTC Classic Trials Championship was initiated and immediately attracted more than 100 registered contenders. As well as the three MCC classics, the Championship includes up to a dozen other trials, among them the VW Owners Club Clee Hills (first held in 1980 and now the VWOC's premier sporting event), the Tamar Trial organised by Launceston & North Cornwall Motor Club, Camel Vale Motor Club's Camel Classic and Woolbridge Motor Club's Olds Classic.

In the first year of the ACTC Championship, Keith Vipond (1300 Beetle) won the Crackington Cup, based on points for class placings; and Ian Facey's 1776cc Beetle was the highest-placed production car in the Wheelspin Trophy League, which is based on overall positions.

In 1986 the VW Buggies of Richard Penhale and Dean Vowden dominated the ACTC Championship, finishing first and second on the Wheelspin League. Vowden went on to win the league in '89, his Buggy starting the season with a Type 3 engine and finishing with a Type 4. David Alderson headed the Crackington League in 1989 with his 1584cc VW Shorty — a short wheelbase Buggy chassis fitted with a shortened Beetle sedan body. Runner-up was Dennis Greenslade in a 1300 Beetle. Greenslade's Beetle was again the highest-placed production car in 1990, and in 1991 the Wheelspin League champion was Arthur Vowden in his unique D'Arth VW Special.

Other VW drivers who have figured in the top six in the ACTC Championships are Grahame Marshall (1600 Buggy), Mark Smith (1300 Baja), Nick Ryle (1776cc 1302S Beetle), Rob Clough (1300 Beetle), Roger Ugalde (1835cc Beetle), Jason Collins (1600 Baja) and Dean Vowden in a Notchback.

One of the fascinations of trialling is that there is no single formula for success. Alan Foster favours an SU carburettor on his 1600cc GT Beetle. He uses a 1300 gearbox and retains the swing axle suspension, with the addition of Konis and coil-over springs at the rear. With this set-up, Foster scored the only outright win by a production car on Ross & District Motor Club's Kyrle Trophy Trial, in 1983. He won Chester Motor Club's Lady Mary Grosvenor Trial outright, has gained two MCC Triple Awards, and was runner-up in the Classic Trials Championship in 1982.

Ken Green has won three MCC Triples and taken premier awards on the Cotswold Clouds Trial (Stroud & District Motor Club) and Chase Clouds Trial (Shenstone & District Motor Club), using a 1300 Beetle with double-jointed rear suspension (USA-spec). He tuned the engine for optimum torque at low rpm by fitting the camshaft and Solex carb from an early 1200. Green's latest trials Beetle again has double-jointed rear and torsion bar front. It's a 1968 semi-auto chassis with a 1302 manual gearbox and fitted with a '72 1200 bodyshell, together with a few parts acquired cheaply from a '66 insurance write-off. The engine is 1776cc with a 1200 cam and Solex 34 PICT carb. Other features are front axle supports, Trekker stub axles, Bilstein shocks, Type 2 bump stops and gearbox mounts, Type 3 clutch, and rear torsion bars raised by two splines.

Hans Viertel's collection of about 100 trophies includes 10 first class awards on MCC trials and 20 class wins on ACTC Championship rounds in Devon and Cornwall. Viertel prefers a 1302S Beetle. He raises the ground clearance as much as possible, but keeps the springing soft for optimum articulation and weight transfer. (He does not reveal the secrets of how this is achieved!) He cuts off jacking points and lower parts of the wings to avoid any drag from the underside, fits wheel spacers at the front, cuts back the steering limit stops to give a tighter turning circle, and uses the lowest ratio gearbox (1300/1302). For maximum torque, Viertel recommends using single port heads for an engine up to 1600cc. He places great emphasis on tyre pressures, saying there is an optimum (within +/- 1 psi) for every different trials hill and every different car, and only experience will get it right.

Dave Keat's 1303S has an 1835cc engine with twin Webers — and a sticker proclaiming 'I gotta lotta throttle'. While other trials drivers say the only way to maintain traction is to trickle along on a gentle throttle using low speed torque, Keat uses full power and full revs, burning the grass and displacing half a ton of mud through wheelspin. Those who try this technique for themselves invariably fail, but Keat makes it work, and, as he says, it's more fun for the spectators! He's made overall best performance on the Clee Hills Trial, beaten all the sports cars and specials on the Camel Classic, and is a regular class winner on the Land's End and Exeter Trials.

On the Tamar Trial, Keat has equalled the scores achieved by the top trials specials such as Mike Chatwin's Dellow and Chris Reeson's Troll, and his Beetle was the first saloon car to reach the top of the notorious, near-vertical Hustyn Hill (near Wadebridge). He has conquered the steep, twisting, slippery section known as 'Park Impossible' at Lifton, near Launceston, and he climbed Crackington Hill (a section on the Land's End Trial between Bude and Boscastle) in second gear on one occasion when it was so difficult that the number of cars getting stuck had caused a three-hour delay.

The most successful VW driver in autotests since 1980 is Graham Hoare, from Dorset, who has won the BTRDA All-Rounders Championship nine times. This Championship takes drivers' results from three different types of motorsport. Hoare competes in autotests, autocross and production car trials. It is perhaps surprising that he uses a Golf or an Audi 80 in trials (very successfully), but he drives Beetles in autocross and autotests, using the same two Beetles for all his Championship victories since 1980.

Hoare's autocross Beetle has a 2.1-litre engine, with twin Weber 48 IDA carbs, dry sump system, and Porsche five-speed gearbox. Hoare can win his class and score FTDs in autocross, but he uses these results primarily as a contribution towards the All-Rounders title rather than contesting the autocross championship. His main strength lies in autotests where he has held the BTRDA Class Championship for many years. He has also won the BTRDA Silver Star and taken third and fourth places overall in the RAC National Autotests Championship.

In autotests Hoare uses a 1776cc 1303S. To attain optimum power and torque characteristics, he uses a single progressive choke Weber carb and fairly mild cam. To facilitate spin turns, he installed a limited slip diff, 13-inch wheels, and lowered the front by about two inches. Koni shock absorbers are still performing well at 14 years old! Hoare modifies the gearshift gate to block off third and fourth gears.

Historic rallying was born in 1982 when the RAC celebrated the 50th anniversary of the RAC Rally by staging the Lombard Golden Fifty. The cut-off date for eligible cars was 1968 — primarily to exclude Ford Escorts, still competitive in contemporary rallying — and entries for the Golden Fifty ranged from a 1927 AC Montlhery Sports to ex-works Mini-Coopers and Austin Healey 3000s. Two VWs took part, Bill Bengry's 1960 model (basically the car he drove on the 1960 RAC Rally, but somewhat modified) and David Lucas' 1957 1200 which he regularly used on trials and autotests, fitted with twin Solexes and Konis. Lucas finished third in the 1940-1960 class. The rally included autotests (where Lucas lost time to the class-winning Sprites) and tarmac special stages.

The Welsh Association of Motor Clubs celebrated its silver jubilee later the same year with a historic rally consisting of special stages over Epynt ranges, and autotests. Ken Green drove a 1300 Beetle on this event and drove a Notchback on a similar rally in 1983.

Subsequently the Historic Rally Register was formed and the movement rapidly gathered momentum. There is now a full calendar of historic rallies, both road navigation events with regularity sections and autotests, and full-blooded special stage events. To date, the only regular VW entrant in historic rallies in the UK is Bob Beales, who drives a '58 Beetle once owned by Bill Bengry. The car was also owned for a time by a taxi cab firm in Hereford who rolled it, and later on in its incident-packed life a garage mechanic drove it into collision with a bus.

Beales acquired the Beetle in the late '60s and used it extensively in rallies, autotests and autocross. He used various engines up to 2.1 litres and also fitted Type 3 disc brakes. The car was then put in storage until 1989, when Beales became interested in historic rallying. He drove it on the historic event supporting the 1989 RAC Rally and finished 32nd.

In the 1990s, with an Okrasa engine, Beales would start registering some major successes, and the ranks of historic VW rallyists would be swelled by Mike Hinde and Francis Tuthill.

The 1980s saw some spectacular developments in rallycross with the advent of turbochargers and four-wheel drive. Audi quattros won the European Rallycross Championship in 1982 (Franz Wurz) and 1983 (Olle Arnesson), and a quattro (Arnesson again) won the Grand Prix in '83. By 1984 there was a whole new breed of rallycross supercars out to beat the Audis. There were Ford Escorts, BMWs, Porsches, Volvos, and, yes, even Beetles — all with specially engineered 4wd systems, megapowerful turbos and massive intercoolers. Then there were the factory-built 4wd Group B rally cars — Ford RS200, Peugeot 205 T16 and Metro 6R4.

In the case of Porsche, power numbers of more than 700bhp were claimed. Acceleration times of 0-60mph in less than 2.5 seconds put rallycross cars up with Formula 1 and drag racers… but the rallycrossers could do it in the mud! In 1985, when Matti Alamaki won the European title, his 4wd 3.2-litre twin turbo Porsche 930 was said to have 900bhp and to go from 0 to 60mph in 2.1 seconds.

Two Beetles were competitive at this level, both with turbochargers, huge intercoolers, overhead cams, and home-made four-wheel-drive systems. Both came from Sweden and their drivers were Mikael Nordstrom and Orjan (pronounced Orryan) Wahlund. In 1985 Nordstrom took fifth place in the European Championship, having taken his Beetle into third place in some qualifying rounds. In view of the power of the opposition, this can be regarded as just as great an achievement as the

Championship wins by Wurz and Teurlings in the '70s.

There were other Scandinavian drivers in 4wd Beetles in the mid-'80s, including Thor Holm and Kjetil Aaen. The first British competitor to use a 4wd Beetle in rallycross was John Aitkenhead, in 1987. He used a turbocharged water-cooled Type 2 engine, but was dogged by problems with the electronic fuel injection. At one televised meeting Aitkenhead had been third fastest overall, but when a TV camera was installed in his car for the final run, it refused to start!

Peter Harrold's PPJ Racing Team/ Autocavan Beetle went to 4wd in 1989. Previously, Harrold had gained an impressive fourth place overall in a round of the Lydden Winter Series, with 2wd. At that time, his 2.1-litre turbocharged, fuel injection, Type 1 engine produced about 400 bhp.

VW drag racing got under way in Britain in 1987 when the first Bug Jam was held at Santa Pod, sponsored by *Custom Car* magazine. About 100 VWs took part, on a 'Run What You Brung' basis. Bernie Smith's full-race two-litre chopped-roof 'Humbug' was quickest, posting times in the low 12-second bracket. John Brewster made fastest time by a street-legal Beetle with 16.2 seconds.

Later that year the VW Drag Racing Club was formed and a new magazine, *Volksworld*, was launched, catering primarily for the customising and drag racing sector. The Volksworld Spring Meeting in 1988 marked the beginning of the VWDRC Championships, in which three classes were established: modified, street, and buggies/kit cars. There were three meetings that year, attracting about 35 competitors. Bob Wick and Keith Seume took the Modified and Street titles, respectively. Seume, editor of *Volksworld*, was running a two-litre Fastback.

In 1989 there were five VWDRC events, using Avon Park and North Weald, as well as Santa Pod. The Modified champion was Luke Theochari, running a 2.3-litre prepared by Terry's Beetle Services; Superstreet champion was John Brewster, in a 2.1-litre Autocavan 1303.

Brewster's car, known as 'Street Lethal', held the record as the fastest street-legal VW in Britain (12.27 seconds/116mph) until 1991. The Autocavan-built engine used Bugpack crank and rods, Scat split-port heads, Scat cam with 1.25 ratio rockers, twin Dellorto carburettors and nitrous oxide system. Maximum power at the wheels (with nitrous) was 245bhp. Chassis components included swing axle rear suspension with adjustable spring plates and KYB shocks, Type 1 'Rhino' gearbox, a billet super diff, and Golf GTi front struts. For this class, cars had to be totally street-legal, with tax, insurance and MOT. The only exceptions allowed were an open exhaust and slick tyres.

In America the Superstreet class is a slight misnomer since it does not actually require cars to be street-legal. On the other hand, it does not permit nitrous oxide or turbos. The VW floorpan, VW-style suspension and VW crankcase must be used. Pro Stock (known as Pro Sedan in the 1980s) allows a tubular space frame instead of the Beetle floorpan, virtually unlimited engine mods ('Stock' is another misnomer!), but again no turbo or nitrous. Pro Turbo, the only class with a meaningful name, does allow turbos.

By the end of the '80s the state of the art in US VW drag racing was being demonstrated by Pat Nelsen's Pro Turbo class winner. The engine was built by Dave Kawell, one of the leaders in VW turbocharging technology, and produced 440bhp from two litres. Nelsen's Beetle ran the quarter-mile in under 10 seconds, topping 140mph. Front runners in the

Pro Stock class, capable of times just over 10 seconds, were Jack Sacchette (Pro Stock champion), Mitch Evenson and Adam Wik.

'The world's fastest VW' is a description that has been applied to the class A dragster built by Dave Folts and Tom Schuh, although in truth it probably contains no actual Volkswagen parts. The rules for this class require a 'VW style' engine; but with crankcase, cylinders and heads manufactured by Autocraft giving a displacement of 2783cc, the only similarity with a Volkswagen would seem to be that it is an air-cooled flat four! With an Airesearch turbo, Hillborn fuel injection, and running on alcohol and nitrous, the power output is more than 600bhp. It drives through a Chevrolet Powerglide transmission. In 1988 Dave Folts set the records at 7.35 seconds and 180.36mph.

All-Beetle circuit racing started in Germany in 1989. Known as the Käfer Cup, it is organised by Käfer Motorsport and sponsored by the German VW-Audi specialist magazine, *Gute Fahrt*. The Käfer cup is not limited to circuit races alone: its nine rounds are divided equally between circuit races, hillclimbs and slaloms. For the first year there were four classes: 1) standard 44bhp, 2) standard 50bhp, 3) modified Type 1, and 4) modified Type 4.

Winner of the Käfer Cup in 1989 was Walter Schäfer, driving a class 3 1302S. Willi Herbig, also running a class 3 car but with a swing axle chassis, was second. The class 4 winner was Kurt Hassmann, who finished third in the overall points standings. 'Fast Edy' Henckel headed class 1 and Gerd Aldekamp was the class 2 leader.

Walter Schäfer had been competing in slaloms and hillclimbs with a Beetle for many years before the Käfer Cup, having started racing his '72 model 1302S in 1977. After carrying out some modifications to increase the car's power for street use, he visited a slalom race in a neighbouring village and decided to have a go. At first, he was regarded as something of a clown to be racing a Beetle, but he soon started winning — not just his 1600cc class, either, but overall!

During the 1980s Schäfer took part in some 200 events in southern Germany, mainly slaloms, but also hillclimbs, and a few races at Hockenheimring and Salzburgring. On most occasions he was the only Beetle driver, competing against 16-valve Golfs, BMWs, and even a Renault Alpine and Lotus Super Seven. Schäfer's best results were second places in hillclimbs at Auerberg in 1986 and Samerberg and Wallberg in 1987. He also won the Beetle drag race at Giebelstadt, near Würzburg, in 1987.

Two other drivers racing Beetles in Germany were Dr Josef Gerold, who drove a 1302S, switched to BMW in the German Touring Car Championship, but is now back in a Beetle for the Käfer Cup; and Josef Plank, who was very fast in a 1500 Beetle.

# John Maher Racing

- Performance Engines
- Close Ratio Transmissions
- Race Car Prep. ● Line Boring
- Case & Head Machining
- Performance Valve Jobs
- Balancing ● 8-Dowelling etc.
- Tube Chassis ● Narrowed Front Axles ● Ladder Bar – Coil Over Rear Ends ● Aluminium Panels
- Lexan Windows

**SERVICE ● REPAIRS ● PARTS**

Call 061-881 5225 or send £1.00 for info pack
Unit 16 Albany Trading Est., Albany Rd., Chorlton, Manchester M21 1AZ.

## Kingfisher Kustoms
### VOLKSWAGEN SPECIALISTS
Telephone:- 021 558 9135

**EVERYTHING FOR THE ENTHUSIAST**

FROM SERVICE TO RESTORATION + PARTS
+ EXHAUSTS + CLUTCHES + WHEELS + TYRES +
+ MOT PREPARATION + ELECTRICS + BODY PANELS
+ STANDARD BEETLE, BAJA AND KAL KITS +
OUR OWN KOMBAT BUGGY + THE CHENOWTH
THE LEADING OFF-ROAD RACING FRAME

**CHECK OUT OUR PRICES** Telephone
# 021 558 9135

UNIT 22
MORNINGTON ROAD
SMETHWICK
WARLEY
WEST MIDLANDS
B66 2JE

## VW AIR COOLED SPECIALISTS
*FROM REPLACEMENT PARTS TO HIGH PERFORMANCE, FROM THE BASIC TO THE TRICK, WE GIVE PERSONAL SERVICE AT ALL TIMES.*

---

### Allshots Beetle Centre
Established 30 years

Allshots Farm, Woodhouse Lane
Kelvedon, Essex CO5 9DF
Tel: (0376) 83295

**LARGE SELECTION OF NEW & S/HAND PARTS ALWAYS AVAILABLE**

We specialise in MoT repairs/welding, crash work, resprays and insurance work undertaken, conversions and restorations.
Suspensions lowered, customising & CAL look
Beetles always wanted – any condition considered.
Established 30 years

---

## FORM + FUNCTION Racing

**FOR ALL AIR COOLED VW STOCK AND PERFORMANCE WORK**

ENGINES, TRANSMISSIONS, SUSPENSIONS, BODY WORK INCLUDING G.R.P. PANELS, SAFETY EQUIPMENT, ELECTRICS, TURN KEY RACE CARS BUILT

THE OLD BOILER HOUSE, KEIGHLEY BUSINESS CENTRE, SOUTH ST., KEIGHLEY, WEST YORKS.
**TEL: 0535 690702**

---

## THE BEETLE SANCTUARY
WELLS-NEXT-THE-SEA, NORFOLK

**Restoration Specialists**
*also* Sales, Service and Repairs.
Spares and Accessories.
ENGINE TUNING
RACE & RALLY PREPARATION
Full suspension lowering facilities.

SPARES PRICE LIST AVAILABLE
LET US PUT A SMILE
ON YOUR BEETLE'S FACE
Phone: Fakenham (0328) 710221

---

## Hoods Galore UK

The Best Original Quality Soft Tops at Wholesale Prices

**Nationwide and Full Fitting available**

**TEL: 081-654 9290**

**FAX: 081-656 9131**

---

## STERLING GARAGE

THE BEST FOR YOUR BEETLE

UNIT 9
STERLING ESTATE
KINGS ROAD
NEWBURY
BERKS RG14 5RQ

(see Dream Machine advert for map)

- COMPLETE OR PART RESTORATIONS
- CAL LOOK MODIFICATIONS
- MECHANICAL SERVICING & MAINTENANCE
- 1K-2K TOP QUALITY RESPRAYS
- BODYSHOP & WELDING WORK
- MOT TESTING AND REPAIRS
- CABRIOLET CONVERSIONS
- MILD TO WILD CUSTOMISING
- URO & DREAM MACHINE OUTLET

TELEPHONE KALVIN OR DAVID FOR FREE ESTIMATES
**(0635) 35721 or (0635) 528953**

OVER 100 YEARS COMBINED AIRCOOLED EXPERIENCE
DISCOUNTS FOR CLUB MEMBERS (PHONE KALVIN FOR DETAILS)

---

## Volksfolk
*people who care*

*Volkswagen Specialists*

*Aircooled Renovation
Lowering. Servicing. Welding
Spraying. Building to customers spec.
Cal look. MOT preparation etc*

Cowbridge, Boston Lincs
**0205 367565**

---

## SCOT VOLKS LTD
VOLKSPARES AGENT FOR SCOTLAND & NORTH OF ENGLAND

☆ VW SPECIALISTS ☆
☆ MECHANICAL REPAIRS ☆
☆ SERVICING ☆

*Discount Card Available*

**FOR ALL YOUR VW PARTS AT DISCOUNTED PRICES**
**CALL US ON 031-337 9500**
(mail order 24-48 hour delivery)

We buy, sell and repair VW Beetles
and can supply all VW Beetle parts
**SAVE MONEY AND TRY US FIRST**
24 Angle Park Terrace, Edinburgh
Tel & Fax: 031 337 9500

# VW BEETLE IN MOTORSPORT

**Left:** *Porsches were the cars to beat in rallycross during the 1980s. Here Karsten Bikskrud from Norway has his Turbo Beetle ahead of Seppo Niittymaki's twin-turbo (700bhp) Porsche. Niittymaki, from Finland, was runner-up in the 1983 European Championship, and he did overtake Bikskrud here at Brands Hatch!*

**Below:** *Most successful VW driver in Belgium was Francois Monten who won the Belgian Rallycross Championship at least seven times in a Beetle. Monten's Turbo Beetle is seen here leading John Greasley's Porsche at Lydden in 1982.*

**Bottom:** *It was believed at the time that this would be the very last appearance by a Beetle in a World Championship rally because the homologation expired at the end of 1983. The rally was the 1983 Lombard RAC, where Austrians Ernst Harrach and Harald Gottlieb entered this 1302S, only to retire shortly after being photographed here on the first special stage at Longleat. In fact, the Brazilian-manufactured Beetle was homologated for motorsport after this and the very, very last appearance in a World Championship rally would be in the 1990s.*

### Opposite page

**Top:** *Francis Tuthill, who has a rally preparation and restoration workshop near Banbury, rallied two Beetles. The 'Rainbow Beetle' is the 1302S he drove on the London to Sydney Marathon, and he also ran it in the Clubman's event following the RAC Rally in 1984, as a charity fund-raiser. Tuthill's second Beetle was powered by an ex-Porsche-Austria 2.4-litre engine and was seriously competitive in National special stage rallies. (F. Tuthill)*

**Bottom:** *Tuthill's 2.4-litre Type 4-engined Beetle in action on the Halewood National Rally. Tuthill regularly finished in the top 20 on Autosport National Championship rallies. When he attained 14th place, then 12th, then eighth, leading Ford Escort drivers were asking worriedly how much faster the Beetle was going to get! The best result for Tuthill was in fact fourth overall on the Peter Russek Memorial National Rally, in 1980. As well as Bilstein suspension with Type 3 torsion bars and limited slip differential, Tuthill's developments included four-pot brakes, with variable balance, and Kugelfischer injection. (C. Money)*

THE 1980s

# VW BEETLE IN MOTORSPORT

*Left:* The start of historic rallying was the RAC Lombard Golden Fifty in 1982. Bill Bengry entered the Beetle he had rallied so successfully in 1960 but the car was not quite original, having a 1500 engine and four-bolt rear wheels! (K. Green)

*Left:* A second historic rally in 1982 was the Welsh Association of Motor Clubs' Silver Jubilee. Ken Green and David Lucas entered this 1300 Beetle, pictured here on a special stage on Epynt, in mid-Wales. (Speedsports)

*Below:* They call this the 'Boss Beetle'. Originally built and raced by Mick Hill in the 1970s, this 'Beetle' has a McLaren-designed (Trojan) Formula 5000 chassis and a five-litre Chevrolet V8 engine. In the early 1980s, driven by Jeff Wilson and with sponsorship from Autocavan, the Boss Beetle was a frequent race-winner.

## THE 1980s

*Above:* Some 30 years after first driving a VW in autocross, Laurie Manifold was still winning. This is his 1980s Beetle, sponsored by Sheppards of Bishops Stortford and powered by a 2230cc Autocavan-built engine with Weber 48 IDA carbs producing some 150bhp. The specification includes close ratio gears, limited slip diff, 13-inch wheels and Bilstein rally shock absorbers that were still going strong after 10 years' use in autocross. The car was classed as a special because the engine was enlarged by more than the +500cc limit for the saloon class, but this did not bother Manifold who was in the habit of making outright fastest time!

*Right:* Alan Foster climbing Litton Slack, in Derbyshire, on the 1986 MCC Edinburgh Trial in his SU-carbed Beetle. Although quite easy in the dry, this section is a notorious 'stopper' in the wet; in 1982 Foster's Beetle was the only production car to reach the summit. Foster has won two MCC Triple Awards; he would probably have won more, but did not compete on the Exeter for several years when he was a member of the organising committee.

*Right, bottom:* Ken Green's trials Beetle in the 1980s was this '69 model lhd American-specification 1300 with double-jointed rear suspension. Green is shown here on the section known as Crooked Mustard, on Stroud and District Motor Club's Cotswold Clouds Trial. In 1981, his Beetle was the first saloon car ever to climb this section, which had been used in trials for 50 years. Green won MCC Triple Awards in 1979, 1980 and 1982, and won the VWOC Trials Championship twice.

## VW BEETLE IN MOTORSPORT

**Left:** *Cornishman Hans Viertel negotiating a section known as Lymer Rake on the Edinburgh Trial in his 1302S. Viertel won MCC Triples in 1985 and 1986, but regards trials in his local Cornwall region to be more challenging. He has gained more than 20 class wins on these events.*

**Below:** *Dave Keat powering his 1835cc 1303S up Simms Hill on the Exeter Trial in 1985. Keat's motto is 'if in doubt — flat out' and his full throttle technique gets his Beetle to the top of hills, through mud and over obstacles which stop most other cars. He is always a leading contender for a class win or outright best performance, but rarely goes after championships due to his work commitments as a farmer in Cornwall.*

## THE 1980s

*Above:* Dave Keat (right) replacing the air in his tyres after a section on the Clee Hills Trial. Keat runs ultra-low pressures and is noted for carrying four spare wheels — partly as ballast but also to cover the likelihood of multiple punctures!

*Right:* Richard Penhale showing what can be done with a Type 3 Notchback. This section would be 'impossible' even without the snow! Powered by a 1776cc engine with twin Webers, Penhale's Notchback has won three MCC Triples, while performances on other classic trials lead one to believe that 'anything a Beetle can do, a Notchback can do better!'

*Right:* Paul Fairbanks calls his VW-engined trials Buggy 'Impetus'. VW suspension is attached to a tubular frame with aluminium panelling. Note the gull-wing doors. The engine is a 1500 Type 3. Fairbanks won MCC Triples in 1978, 1980, 1982, 1983 and 1984, and won the Baddeley Award for overall best performance in MCC trials in 1982.

# VW BEETLE IN MOTORSPORT

*Left:* This vehicle has not just been in an accident! It's Peter Tonks' Bradbury Special, another tubular-framed trials Buggy powered by a Type 3 VW engine. The doors are detachable and carried where they can contribute to rear weight distribution.

*Below:* A picture to gladden the hearts of all Beetle motorsport enthusiasts! Two Beetles, driven by Swedes Mikael Nordstrom and Orjan Wahlund, lead the field in rallycross at Brands Hatch in 1985. By this time all the top rallycross cars, including these Beetles, had turbos and four-wheel drive.

*Bottom:* Probably the ultimate Beetle racing engine. This is Orjan Wahlund's turbo-charged, overhead-cam, 2.1-litre, which developed about 450bhp. (And, unlike similarly powerful drag race engines, this one ran many miles round twisty, bumpy rallycross circuits in mud, rain and dust!) The tubular cage holds the power unit and car together and protects the engine in the event of a rear end collision.

*Opposite page*

*Top:* A massive intercooler for the turbo on Wahlund's Beetle is incorporated in a whale-tail spoiler.

*Bottom:* Mikael Nordstrom's 4wd Turbo Beetle in action at Lydden in 1985. Nordstrom took fifth place overall in the European Championship. This Beetle was well capable of racing with Le Mans-engined Porsches, Audi quattros and the formidable Ford Escort Turbo X-Trac of Martin Schanche.

THE 1980s

*Left: The engine of Nordstrom's rallycross Beetle. The specification is similar to Wahlund's but the turbo is located in front of the engine, above the gearbox, instead of behind.*

*Left: The Nordstrom Beetle's 4wd system uses a Porsche gearbox driving to a BMW differential at the front, with Passat drive shafts, Golf struts, and a SAAB freewheel.*

*Below: Another 4wd Turbo Beetle, rallycrossed in 1985 by Thor Holm. This photograph shows the drive to overhead camshafts using Golf components.*

THE 1980s

*Above:* Keith Vipond at Bluehills Mine on the Land's End Trial in 1984. Vipond won a Triple Award that year and won the ACTC Classic Trials Championship in his 1300 Beetle.

*Right:* Dean Vowden starting trialling a Buggy in 1985 when he was only 18. He scored outright wins on the Tamar Trial, the Wessex Trial and Holsworthy Motor Club's Taw and Torridge Trial. Vowden was runner-up to Richard Penhale in the 1986 ACTC Championship and he gained the Austin Hannam Cup for overall best performance on the Exeter Trial in 1987. In this shot from the 1986 Exeter, Andy Rice is acting as look-out from the back seat (!) while Vowden drives up Fingle Hill.

VW BEETLE IN MOTORSPORT

***Above:*** *Dean Vowden's second Buggy was initially powered by a 1600cc Type 3 engine and later by a two-litre Type 4 engine. Vowden became the overall ACTC Classic Trials Champion driving this Buggy in 1989. Here he is competing on the VWOC Dhustone Trial which took place in and around an old quarry on the top of Clee Hill.*

***Opposite page. Top:*** *Ian Facey's 1776cc Beetle was the highest-placed production car in the 1984 ACTC Championship. Facey has scored numerous class wins, including the Clee Hills (shown here in 1987) when he finished second overall and again was best production car.*

***Opposite page. Bottom:*** *Winner of the 1987 Clee Hills Trial was a 1584cc Buggy driven by Richard Penhale. Following his successes with the Notchback, Penhale switched to a Buggy and became overall ACTC Champion in 1986. Here, assisted by the 'bouncing' of Kate Taylor, he is starting his non-stop climb of the Jenny-Wind section. They were the only crew to reach the summit.*

THE 1980s

## VW BEETLE IN MOTORSPORT

*Left:* David Alderson won the ACTC Crackington League in 1989 with his VW Shorty, a vehicle combining the features of a short wheelbase Buggy and a Baja Beetle. Shorty is in action here on the Bristol Motor Cycle & Light Car Club's Allen Trophy Trial at Big Uplands.

*Left:* A scene from the 1986 Rallycross Grand Prix. Orjan Wahlund's 4wd Turbo Beetle leads Will Gollop's Metro 6R4, Andy Bentza's Audi quattro, Piet Dam's 4wd BMW Turbo and Bengt Viklund's Ford RS200. Supercars all!

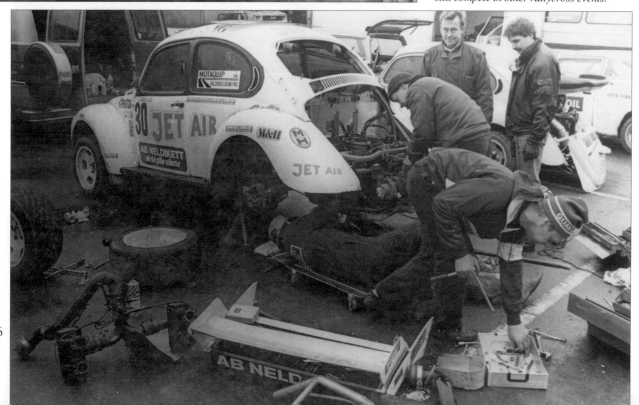

*Below:* Just a little job in the paddock between races as Wahlund and crew replace the gearbox mounting on the Turbo Beetle at the Rallycross Grand Prix in 1987. Wahlund qualified for the 'C' final, in which he finished fifth. By 1987, due to expiry of the homologation, Beetles were no longer eligible for European Championship rallycross, but they can still compete in other rallycross events.

THE 1980s

*Right:* John (Acerace) Aitkenhead built this superb 4wd Beetle for rallycross in 1987. The engine is a turbocharged 2.1-litre water-cooled flat-four Type 2, giving around 300bhp, and it drives through a modified Transporter five-speed gearbox and home-made prop shaft to a Lotus Elan differential at the front. Audi 80 suspension and Golf brakes are used. Rear suspension is basically 1303 Beetle, but with coil springs instead of torsion bars. Problems with the electronic fuel injection prevented the car from achieving its full potential, its best result being third fastest in a Shell Oils Rallycross at Brands Hatch. Lack of sponsorship and the recession prevented further development, though Aitkenhead hopes to rallycross the Beetle again in 1993.

*Right:* Australian Richard Hölzl likes to race ovals — not so much oval circuits as oval-window VWs. This was his first, a '55 body on a '69 semi-auto chassis with double-jointed rear, converted to manual transmission and powered by a two-litre engine. Hölzl used the car mainly in super-sprints which were two or three timed laps of a circuit, with cars running in pairs. He made several outright fastest times. Hölzl also drove this Beetle in his first motor race, at Amaroo Park, Sydney, which he won, finishing half a lap ahead of the field of 26 which included Ford Falcon GTs and V6 Alfa Romeos. The Beetle remained a road car and covered no less than 235,000km in five years! (R. Hölzl)

*Right:* Classic trials live out the old slogan once used in VW advertising: 'A VW goes even where there is no road'. This 'unsuitable for motor vehicles but ideal for Beetles' track is Lymer Rake, in Upper Dovedale, Staffordshire. The Beetle is a 1200 driven by M. Bradley on the Edinburgh Trial.

VW BEETLE IN MOTORSPORT

**Left:** *Another Cornish trials ace is Jon Robilliard, who enhanced the weight distribution of his Beetle by mounting no less than three spare wheels behind the engine. He has scored class wins on the Lands End and Exeter classics and on many other trials in Devon and Cornwall.*

**Below:** *Simon Woodall and Barbara Selkirk reaching the top of a section on the Dhustone Trial in their Buggy. Woodall won this event outright in 1983.*

# THE 1980s

*Above: Alan Bellamy uses a Type 3 Fastback in both classic trials and production car trials. Here he is on his way to overall best performance by a saloon car on the Dhustone Trial in 1988.*

VW BEETLE IN MOTORSPORT

**Above:** Grahame Marshall regularly took his Buggy into the top five of the ACTC Championship. He is in action here on the Woolbridge Motor Club's Olds Motor Group Classic Trial, climbing a section on the military vehicle testing ground at Bovington Camp. Marshall finished second overall on this event, beaten only on a tie-decider by Chris Reeson's Troll.

**Left:** Variously known as VW model 181, 'Trekker' and 'Thing', Volkswagen's factory-built off-road buggy does not really offer any advantages over a Beetle for trialling. However, this example has competed in trials, driven by Mark Tooth, who subsequently changed to a Beetle in which he won an MCC Triple Award in 1992.

## THE 1980s

*Above:* John Parsons was the first to use an off-road 'racer' in trials. This is a Fugitive, powered by a 1700cc Type 4 engine. Parsons has had some success, including several class wins, though the long wheelbase is a handicap for trialling.

*Right:* Because of their bodywork modifications, Baja Bugs are classified with kit cars and limited production cars, such as the Marlin, and may have to carry out more difficult restart tests than the Beetles. Steve Dorrell, who has won many awards with his Baja, is seen here successfully clearing a restart line on the Exeter Trial's Tillerton section. The line was positioned close to a step in the rocky surface and many competitors failed.

*Right, below:* Jim Calvert, the Proprietor of Stateside Tuning in Middlesex, ran a Type 4-engined 1303S in British National special stage rallies where his best result was a class win on the Tour of Cumbria in 1984. Calvert then participated regularly in European Championship and West Euro Cup rallies where his Beetle, although never among the winners, drew a great deal of attention competing against Audi quattros, Opel Manta 400s and Porsches. Pictured here in parc fermé during the 1988 Köln Ahrweiler Rally, the Beetle was powered by a 2.5-litre engine with twin Dellorto 48mm carbs developing 200bhp. Other features included a dry sump system, rear disc brakes and Leda adjustable suspension. On a straight at the Nürburgring, Calvert was timed at 126mph.

# The 1990s

## *The 150mph Beetles*

It is surprising that drag racing should be a popular form of motorsport for Beetles: the quarter-mile straight-line tarmac sprint is all about ultimate power, which was never one of the Beetle's attributes, and it takes no account of the VW's strengths which have always been long-distance, all-weather reliability over rugged terrain, traction on slippery surfaces, and an ability to 'go where there is no road'.

Despite being the automotive world's shortest form of competition, drag racing has produced the fastest VWs. An official land speed record for a VW Beetle was set at Bonneville Salt Flats in 1991 at 132.6mph, but the quickest Beetles in drag racing are regularly achieving around 150mph. This apparent anomaly is explained by two factors. The land speed record car has to be very much closer to the standard production vehicle than a 'Pro Turbo' drag racer, and it has to maintain its speed for several miles. Drag racing vehicles have only to run for a few seconds and would undoubtedly self-destruct if run at maximum power for the length of a land speed record course.

Classes in drag racing bear no relation to classes in other fields of motorsport. And the extent of modifications, up to the point where a car is built purely as a quarter-mile drag racer using specialised components throughout and no original Volkswagen parts at all, makes it difficult to say what should, or should not, be regarded as a Beetle.

Since dragsters mostly have large water-cooled V8 engines, anything with an air-cooled flat four is generally accepted by the VW fraternity. At the top of the performance ladder are single-seat, open-wheeled 'rails' with VW-style engines, running on special fuel, such as the 600+bhp dragster of Dave Folts and Tom Schuh which set new records in 1991 of 7.3 seconds and 182.62mph. In the UK there are dragsters of similar configuration being raced by Bernie Smith and Cliff Watkins. Smith's dragster, named 'Industrial Disease', made its debut in mid-'92 and ran 8.8 seconds and over 161mph. He subsequently made an 8.39 second pass, but was unable to back it up by another run within the prescribed tolerance.

Funny Cars are definitely not a joke. They are seriously fast and come second only to the

THE 1990s

*Above:* Keith Seume's Autocavan-sponsored turbo Beetle, known as 'No Mercy', set the official record for the fastest Beetle in Britain in 1991, covering the quarter-mile at Avon Park Raceway in 10.08 seconds and reaching 138.98mph. The split-window 1952 body has a tubular space-frame chassis and the roof is lowered by three inches.

*Right:* The 2.2-litre turbocharged engine of Seume's Beetle produces 470bhp with nitrous oxide injection. The turbo system was supplied by Kawell Racing Engines, of Tennessee, USA, and uses a 48mm Dellorto carburettor. Rear aerofoil came from a Formula 1 Lola.

open-wheeled dragsters for ultimate performance. A Funny Car is an out-and-out dragster clothed in a thin plastic, one-piece sham bodyshell which loosely resembles a production saloon, usually distorted to make it lower and longer.

Brian Burrows' 'Outrage 2', driven by Paul Hughes, is a Beetle Funny Car, built in Britain but incorporating many US-sourced components including an Autocraft engine. This 'VW style' air-cooled flat four comprises an Autocraft crankcase, Scat crankshaft, Carillo rods, Autocraft pistons, Autocraft cylinder heads, Folts camshaft and intake system, Hillborn injection and a Turbonetics turbocharger. Capacity is 2.8 litres, power output (on methanol) is 682bhp, and the cost of the engine has been quoted as $20,000. It is interesting that drag race engine builders do not go to the Type 4 componentry or overhead cam conversions favoured in circuit racing and rallycross in Europe. American technology sticks with the VW push-rod configuration, although all of the valve gear is specially designed. The engine is in front of the driver and it drives through a modified Chevrolet Powerglide two-speed automatic transmission to a modified Ford 'solid' rear axle. Some Beetle!

'Outrage 2' made its first run in 1992 and Hughes was very soon running under nine seconds and over 150mph. By the end of the season, his best numbers were 8.50 seconds (at Santa Pod, backed up by another run in an identical time) and 161.23mph (at Avon Park). Hughes' 8.50 second ET made 'Outrage 2' the fastest flat-four-cylinder-engined car in Europe at that time.

Describing the driving technique, Brian Burrows says that on the start line the transmission is locked and the engine run at full throttle but rev-limited to 5000rpm by a 'splutter-box' which cuts the ignition, one cylinder at a time, in random sequence. This is activated by a thumb button on the steering wheel: at the green light, the driver has only to move his thumb. This releases the hydraulic transmission lock, switches the splutter-box off, and sets a second rev-limiter at 7800rpm which will electrically activate the transmission shift from first to second gear. The torque converter is specially modified to give a high stall characteristic and to cope with the high temperature generated by locking the transmission at full throttle at the start.

The Beetle Funny Car is very stable at high speed — as confirmed by an incident when the steering wheel came off during a big 'wheelie' and Paul Hughes had to steer by grasping the spindle!

Classes such as the USA Pro Turbo and Pro Stock require some of the original Volkswagen metal to be retained, although not necessarily either the floorpan or the crankcase! The class definitions in drag racing are changed so frequently that any effort to detail the rules would be out of date almost as soon as completed.

Dave Perkins was the first to achieve 150mph in a Beetle, which he did on a 9.63-second run at Phoenix in 1987. At that time the Beetle was classified as an A/Sedan. The car had a VW factory floorpan and full-height Beetle body. The engine was 2.1 litres and produced about 450bhp. The spec included a Scat crankcase, Autocraft crankshaft, Superflow heads, Rajay turbo, Holley Dominator carb, and a liberal injection of nitrous. In 1991 Perkins achieved 9.1 seconds and 155mph, but this was in qualifying and therefore not an official record.

At Sacramento in 1992, running in the Pro Turbo class, Perkins established a record at 9.23 seconds and 148.52mph. This was with a new engine — one of the first to use an intercooler on a turbo drag car.

# THE 1990s

Others who have run under 9.5 seconds in the USA are Ron Townsend and Chris Bubetz. Championship-winner in Pro Turbo in 1991 was Robert Couse, whose best time was 9.71 seconds/144mph, using a 2.3-litre engine with RotoMaster turbo and Holley carb in a Ron Lummus-built tube chassis Beetle.

In the Pro Stock class (no turbos or nitrous), Jack Sacchette's Beetle has probably won more races than any other car, taking the Championship title in 1989, 1990 and 1991. Sacchette set an ET record of 9.82 seconds at Sears Point Raceway in 1991, but the speed record in Pro Stock was taken by Mitch Evensen with 133.36mph in 1991 and then by Jorge Walker with 134.52mph in 1992. Sacchette also used a Lummus tube chassis. His 2.5-litre engine had a Scat case and crank, Autocraft heads with 53mm inlet valves, and Weber carbs opened out to 53mm.

The Super Street class in the USA does not require cars to be street legal, but it does (as of 1992) impose a weight limit of 10.5lb per cubic inch of engine capacity. Gary Berg set the record with a 10.55 second/124 83mph pass at Sacramento in 1992. Built by the Berg family — Gene, Clyde and Gary — the 2.6-litre engine delivers 280bhp. An ARPM crankcase houses a 94mm forged Berg crankshaft, with Carillo rods and 94mm Cima cylinders. A custom-ground camshaft has roller cam followers and the Superflow heads are special castings designed by Clyde Berg. Valve sizes are 48mm and 40mm, compression is 12.7 and carburettors are 48 IDAs opened up to 52mm.

Main contenders for the 'Fastest Beetle in Britain' title are Keith Seume and Luke Theochari. Seume has recorded 10.08 seconds and 138.98mph and is the 1992 Modified Class Champion. His Autocavan-sponsored split-window Beetle has a 2.2-litre turbo engine with RIMCO-prepared VW Type 1 crankcase, Pauter Machine wedge-mated crankshaft, Carillo rods and Kawell Racing Engines heads with 42mm and 37mm valves. The turbo system is also the work of Dave Kawell. TRP Racecars of Tennessee supplied the chrome-moly tube chassis and Harold Carter's Gearbox Shop in Santa Ana built the transmission.

Theochari was the first to break 10 seconds, but was unable to back up his 9.96-second time with the requisite second run within one percent. He has since achieved 10.13 seconds and 134.26mph on an amazing run when the back wheels came off the ground due to a shock absorber problem. Theochari's car is built by Terry's Beetle Services, in west London. The 2.3-litre engine uses a Scat crank, Superflow heads with 46mm and 40mm valves, and Weber 48 IDA carbs. There's no turbo, but lots of nitrous.

The engine actually 'grenaded' (caught fire) in 1991 due to being 'too greedy with the nitrous'. It has now been specifically built to run with nitrous, which means (among other factors) a low compression ratio of 9.5. There are two stages of nitrous, the first being activated by a thumb button on the steering wheel and the second coming in when fourth gear is engaged. As on Burrows' car, there are two rev limiters, the lower limit enabling the throttle to be floored before the clutch is engaged, thereby minimising reaction time.

Theochari was VWDRC Modified Class Champion in 1989 and runner-up in 1990.

The VWDRC Championships are decided by 'bracket racing' which enables slower cars to compete on equal terms. Competitors dial-in their target times (based on qualifying runs) and the faster cars are effectively handicapped by the differences in dial-in times. Winning depends on consistency and reaction time on the start line, rather than maximum horsepower.

The most consistent winner is Paul Miller, Modified Class Champion of 1990 and 1991. Miller's Beetle is built by Microgiant in Rayleigh, Essex. The engine is only 1584cc and is a good example of what can be achieved using stock VW components. The standard crank is tuftrided and eight-doweled, the standard rods are stress-relieved and shot-peened, and the standard heads are modified to take 40mm inlet valves and 35.5mm exhaust valves. The cam is an Engle 130 with 1.25 ratio rockers, carbs are dual 45mm Dellortos, and there's a dry sump oil system using a DDS pump and home-made tank. Power on a rolling road, described by Microgiant's Tony Royston as 'very conservative', is 105bhp at the wheels.

The '63 body is on a '68 semi-auto chassis, partly space-framed, with lots of aluminium and fibreglass. Gearbox is from a 1302, with limited slip diff. Mini wheels with Lambretta tyres are fitted at the front, and modified Skoda wheels with ex-hillclimb slicks at the rear. Best numbers recorded to date are 13.2 seconds and 98.88mph, and there must be a lot of racers with more expensive and exotic cars who would be happy to run that quick!

In the classic trials arena, there have been more successes for the Vowden family, from Torquay. Arthur Vowden and his son Dean designed and built a VW Special which, unlike most vehicles of the genre, actually looks like a seriously competitive trials car, on the lines of a Cannon or Troll. Based on a radically-shortened Beetle floorpan, it is powered by a Type 3 engine with twin Dellortos. Vowden has softened the torsion bar front suspension and added coil-over shocks to give the desired springing characteristics and ground clearance with a very light front end.

Arthur Vowden drove the Special to become 1991 ACTC Classic Trials Champion. His wife Jenny drove it on MCC events, gaining first class awards on the Exeter and Lands End Trials, and only narrowly missed becoming the first female Triple Award winner by stopping on the notorious Litton Slack section in Derbyshire.

Dean Vowden bought a Notchback in 1989 and installed the two-litre Type 4 engine from his Buggy. In 1990 he made overall best performance on the Exeter Trial and gained class wins on the Lands End and Edinburgh to win an MCC Triple plus the Baddeley Trophy and the Team Award (together with the Beetles of Mike Jones and Mark Smith). Vowden's Notchback even beat Dave Keat's powerful Beetle on special test times. He also won awards on the Clee Hills, Chase Clouds, Camel Classic and Cotswold Clouds Trials in 1990. Vowden reckons a Notchback is better for trials than a Beetle, as ballast can be carried in the rear luggage compartment directly above the engine. He changed the rear suspension to double-jointed and modified the steering to improve the turning circle. The Notchback now belongs to Ken Green and Vowden has turned his attention to a space-framed VW-engined Special.

Dennis Greenslade became a 'Triple Triple' winner, achieving nine consecutive unpenalised performances with his 1300cc Beetle in MCC trials in 1990, 1991 and 1992. This was the first time this had been achieved in a saloon car. Other VW drivers gaining Triple Awards in the early '90s were Alan Foster (GT Beetle), David Alderson (Shorty), Alan Bellamy (Fastback), Richard Penhale (Notchback), Jason Collins (1600cc Baja), Mark Tooth (1835cc Beetle) and Robert Clough (1300cc Beetle).

Simon Woodall built a new trials car which made its debut on the Edinburgh Trial in 1992. It is a Baja Convertible with 18-inch ground clearance and 1911cc Type 4 engine with twin

## THE 1990s

Dellortos. Woodall used a '58 chassis with the remains of a 1302S body and a Type 2 rear axle. He says it is the first Baja to be purpose-built for classic trials — and it will be warmer than his Buggy in the winter months!

In historic rallying in 1990 Bob Beales scored class wins on the Welsh Retrospective Rally, the Targa Rusticana and the Clumber Park Stages, and was placed fourth in the Welsh Championship. The following year, with an Okrasa TSV 1300 engine in his '58 Beetle, he won the Welsh Championship outright and was a class winner on the Longleat Stages, Telford Stages, Targa Rusticana (third overall), Bristowe Rally (fifth overall), Circuit of Ireland Retrospective (sixth overall) and Rally of the Vales (fifth overall). Main opposition on these events comes from sports cars: Austin-Healey Sprites, MGBs, Mini-Coopers, Lotus-Cortinas and others.

In '92 Beales tied for an outright win on the Targa Rusticana, but lost to a Mini-Cooper on the tie-breaker — a *concours d'elegance*! For historic rallying in 1993, Beales has prepared a '64 1500S Notchback, similar to the famous Scania Vabis rally cars.

Francis Tuthill (and others) discovered that, while the European Beetle homologation had expired in 1983, the Brazilian Beetle (or Fusca) had been more recently homologated in Group A and this did not expire until the end of 1992. Tuthill set about building a car for the '92 RAC Rally which really would be the final appearance by a Beetle in a World Championship event.

The Beetle built by the Brazilian VW factory in the 1980s was something of a hybrid: it had the 'small window' 1966-style bodyshell, but with the biggest and latest tail-lamps, torsion bar and swing axle suspension with disc brakes, and 1600cc engine. Twin carbs were also homologated.

In Beetle racing in Germany the Käfer Cup classes were changed in 1990 to comprise Division V, which is restricted to the stock Solex 34 PICT carburettor, Division W1 for modified Type 1 engines, and Division W4 for modified Type 4 engines. Catalysts became compulsory in 1992 (Porsche Carrera catalysts are fitted).

There is freedom regarding chassis specification in all classes. Either 1200/1500 or 1302/1303 Beetles, or indeed Type 3, can be used. Swing axles can be, and usually are, replaced by double-jointed rear axles. Torsion bars can be replaced by coil springs. Five-speed gearboxes (Porsche 911 or 914) are permitted, and brakes usually come from the Porsche 944. A survey of the list of race winners since the inception of the Käfer Cup shows a ratio of 67 wins by 1302/1303 to 31 wins by 1200/1500 Beetles.

A typical W1 engine is 2180cc (82mm stroke by 92mm bore) with 44mm inlet and 36mm exhaust valves, 11 to 1 compression and twin Weber 48 IDA carbs. Power is about 150bhp.

W4 engines are up to 2866cc (86mm x 103mm) with 47mm and 39mm valves, 12 to 1 compression and 48 IDA Webers, and deliver up to 220bhp. Dry sump oil systems and Porsche-type cooling are used. Fuel injection and 16 valves are possible future developments. The two principal engine builders are Kurt Hassman, who won the Käfer Cup as a driver in 1990, and Rolf Holzapfel.

Fastest lap times are more than 82mph on the Nürburgring short circuit and 96.95mph round the fast Salzburgring, both by Dr Jo Gerold. The latter time would have put Gerold's Beetle among the top three in the Polo G40 race.

Champions in 1991 were Gerold Probst (Div. V), Walter Schäfer (Div. W1) and Ingo Gaupp (Div. W4). Schäfer was the overall Käfer Cup winner.

Schäfer won the Käfer Cup again in 1992, his third victory in this Championship in four years and a clear demonstration of his all-round ability in the three disciplines: slaloms, hillclimbs and circuit racing. The runner-up and Division W4 winner was Dr Gerold, who won the final race of the season on the Nürburgring Grand Prix circuit. Gerold's fastest lap was 2 min. 01.83 sec. (83.35mph); in qualifying, he did 2 min. 00.15 sec. Ingo Gaupp came second in the race but finished only seventh in the overall Käfer Cup table.

In all, 32 drivers scored points in the 1992 Käfer Cup and it is noteworthy that the top ten places comprised four cars from Division V and three each from Divisions W1 and W4. Coming into the final race, any one of four drivers —Schäfer, Gerold, Ralf Schaub and Klaus Sievers — could have won the title.

Division V was a see-saw contest between Schaub and Sievers. At mid-season Sievers actually led the overall Käfer Cup table, thanks to his excellent performances — despite having only a low-powered 1600cc engine — in slaloms. Schaub then went ahead when he added a class win at the Siegerland circuit to his previous Division V victories at Osnabruck hillclimb and Salzburgring. Then for Nürburgring Sievers fitted a new big engine; he won the class and became Division V Champion.

Beetle racing started in Britain in 1992, promoted by the Big Boys' Toys division of Volkspares and organised in conjunction with the British Racing and Sports Car Club. Cars run in a single class, using sealed engines supplied by Volkspares so that all have the same specification and power output. The engines are 1641cc with a single 40mm Dellorto carb and Engle cam, but otherwise virtually standard. Power is reckoned to be about 70bhp at the wheels. The cars must be swing axle 1200/1300/1500 Beetles. Suspension modifications, wheel and tyre sizes are strictly controlled.

The Beetle Cup is a low-cost form of motor racing which has attracted many first-time drivers. Although speeds are low (the Beetle lap record at Brands Hatch is 69.5mph, compared to the Polo G40's 78.3mph), the racing has proved to be more entertaining, more competitive and with fewer accidents than many critics predicted.

For the first half of the season, the clear winner in every race was Julian Lock, one of the few drivers with previous circuit racing experience. Behind him, however, there was invariably close racing throughout the field. The two drivers to emerge from the pack and challenge Lock's supremacy were rallycrosser John Aitkenhead in his Acerace-prepared car, and Simon Howarth in a car built by Francis Tuthill and backed by Fast Lane Leisure. Both drivers recorded fastest laps. By the end of the season Howarth was winning some races, but the Beetle Cup was already Lock's.

All-Beetle racing is also planned in Australia, though in the meantime VWs are racing very competitively against other makes. Beetle drivers Richard Hölzl, Andrew Frazer and Paul Zanello are regularly finishing in the top six, ahead of 40-plus 'other make' sports sedans. Hölzl's engine is 2.3 litres, but Frazer and Zanello run in the two-litre class and one or the other invariably comes out as class winner. Frazer is headed for a win or runner-up overall in the Bridgestone Club Car Championship, which is for nominally road-registered cars and comprises races at Amaroo Park, Oran Park and Eastern Creek.

The words of Richard Hölzl provide an appropriate conclusion to the story of Beetles in motorsport up to the present time: 'In circuit racing against sports sedans, the Beetle reaches beyond all expectations, and it will go faster and faster beyond the year 2000!'

# ACE AUTO SPARES

## VW / AUDI
## MERCEDES / BMW
### Save £'s £'s
Quality Replacement Parts at Discount Prices

**\* NATIONWIDE MAIL ORDER SERVICE**

230 HIGH ROAD, CHADWELL HEATH, ESSEX RM6 6AP
**TELEPHONE 081-599 4303/8356**
FACSIMILE 081-503 8749

**\* 125 GREENGATE STREET, PLAISTOW E13**
**TELEPHONE 081-470 9782**
FACSIMILE 081-472 0081

---

# RESTORATION PARTS?
*We've got it all!*

## THE LARGEST SELECTION OF VOLKSWAGEN & PORSCHE PARTS IN EUROPE

## KARMANN KONNECTION
mail order hotline
# 0702 551766

60-page restoration catalogue £4 inc p&p.
All prices include VAT. Please add 10% for p&p.
All major credit cards welcome.
**RETAIL SHOP OPEN: 9-6 Mon-Sat. 10-5 Sun.**

4-6 High St, A13, Hadleigh, Essex SS7 2PB. Fax: 0702 559066.

---

**081-317 7333**

WE WILL RESTORE YOUR BEETLE TYPE 2, 3 OR 4
TO THE VERY HIGHEST OF STANDARDS
PAINTWORK WAY BEYOND ANYTHING VAG EVER DREAMED OF

| | |
|---|---|
| CHASSIS | LOWERING |
| WELDING TO THE HIGHEST OF STANDARDS | SERVICING |
| BODY REPAIRS INSURANCE COMPANY APPROVED | ALL MECHANICAL REPAIRS |
| PARTS DEPARTMENT | ELECTRICAL FAUT FINDING SERVICE |
| SPARES NEW & SECONDHAND | ALARMS |
| MAIL ORDER | ELECTRIC WINDOWS |
| FULL MECHANICAL FACILITIES | REMOTE CONTROL DOOR OPENING |
| 081-317 7333 | STOCKIST OF PIONEER/KENWOOD INCAR AUDIO EQUIPMENT |
| | 081-317 7333 |

**081- 317 7333**
MAIL ORDER SECOND HAND PARTS COUNTRYWIDE
UNIT 1, WHITEHART ROAD, PLUMSTEAD, SE18 1DG

---

## Toni's V-DUBS
Beetle specialists
**TEL 0928 580883**

**4 ARKWRIGHT COURT**
ASTMOOR ● RUNCORN
● CHESHIRE ●

---

# BEETLE COSMETICS

- RESTORATION
- CONCOURS PAINTWORK
- CUSTOM PAINTWORK
- STRAIGHT PAINTWORK
- M.O.T. REPAIRS WELDING
- DESEAMING
- LOWERING
- WAXOLY RUST PROOFING SERVICE

THE HIGHEST QUALITY WORK AT PRICES YOU CAN AFFORD, **NO JOB TOO BIG — NO JOB TOO SMALL**

UNIT 4,
LEWISHAM RD.,
INDUSTRIAL ESTATE
SMETHWICK
WEST MIDLANDS
B66 2BP

**TEL:**
**021- 555 - 6807**

---

## LR SUPERBEETLES
### VOLKSWAGEN BEETLE SPECIALISTS

Beetles bought and sold
Repairs ● Renovations ● Spares ● Resprays etc
**Telephone: Colchester 0206 563433**

WE ALSO UNDERTAKE PORSCHE 356 RESTORATIONS
*Specialist in the restoration service and repair of the early Volkswagen*
We offer an extremely competitive service

*Above:* Detachable one-piece front end of Keith Seume's Beetle reveals the JT aluminium front axle beam and Wilwood ultra-light disc brakes. Nitrous bottle is next to the fuel tank.

*Right:* The cockpit of Seume's drag race Beetle. 'Pro-control' boxes enable the settings of the max rev limiter and gearshift light to be adjusted. Dials display fuel pressure, turbo boost, oil pressure and cylinder head temperature. Buttons on the steering wheel activate the front brake hydraulic lock (used when doing the 'burn-out') and over-ride the nitrous switch which normally operates on full throttle. The lever beside the gearshift is effectively a handbrake, used for staging on the start line.

*Below:* Luke Theochari, driving the Terry's Beetle Services 'Moody', recorded a 9.96 seconds/136.6mph quarter-mile at North Weald in 1990, achieving this without a turbo and with a full-height Beetle body and VW floorpan. He was unable to back this up with another run at the same meeting to claim the record, but has since run many times in the low 10-second bracket and expects to go even faster.

THE 1990s

*Above:* Theochari's non-turbo 2.3-litre engine looks surprisingly simple and bare. Carburettors are twin 48IDA Webers and power is boosted by two stages of nitrous. Inside the engine and control system is a great deal of expertise and technology to optimise the nitrous and fuel settings.

*Below:* Jim Bowen started racing in 1989 when he came third in the Street class of the VWDRC Championship with a 1776cc engine. By 1991 the car had been named 'Betelgeuse' and fitted with a two-litre engine and close-ratio gears, whereupon Bowen smashed John Brewster's long-standing Super Street record, became champion of that class, and won a Driver of the Year award. In 1992 he rebuilt and lightened the car and set a new record at 11.24 seconds and 117mph, again winning the Super Street Championship.

*Left:* Bowen tuning his 48IDA Weber carbs. With nitrous oxide injection, the car develops about 250bhp. Bowen acknowledges help from his father and Autocavan's John Brewster.

## VW BEETLE IN MOTORSPORT

**Left:** Jim Warner is a regular competitor in the Street class with his very clean 1776cc Notchback, in action here at the Volksworld Summer Nationals meeting at Avon Park. To be eligible for the Street class, cars must not be capable of running quicker than 15.99 seconds.

**Below:** Gary Angell's 'Bugbear', pictured here at Santa Pod, is another 11-second challenger. Angell regularly battles hard with Bowen for Super Street honours and finished a close second in the Championship table in 1991 and 1992.

**Bottom:** Bill Griffiths making adjustments to the 2.3-litre engine of his 11-second Modified class contender, 'Bad News'.

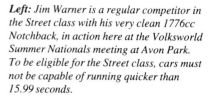

### Opposite page

**Top:** The 1584cc Microgiant Beetle, 'Prince of Darkness', driven by Paul Miller, was VWDRC Modified Class Champion in 1990 and 1991, proving that you do not need radical and expensive engine components (or fancy paintwork) to win in drag racing.

**Bottom:** Jeff Copson's Whitworth Engineering Volksrod was the VWDRC Kit Cars and Buggies Class Champion in 1991. Copson's best run was 12.9 seconds and 98mph, achieved without nitrous. The 2267cc engine is equipped with 45mm Dellortos and Eliminator heads incorporating some 'secret' mods. Other than that, the Volksrod Buggy is remarkably standard and could run as a street legal car — although it has since been reclassified and runs as Modified in VWDRC events.

122

## THE 1990s

## VW BEETLE IN MOTORSPORT

**Left:** *The four-wheel-drive, 16-valve rallycross VW of Peter Harrold made its debut in 1992. It is undoubtedly the most advanced motorsport Beetle built to date. Subaru water-cooled cylinder heads are used, together with Kugelfischer fuel injection and a Garrett turbo, producing more than 500bhp. The 4wd system uses an Autocavan-modified Porsche five-speed gearbox, driving to a BMW differential at the front, with Golf drive shafts. The fully adjustable suspension has coil spring struts and lower wishbones, front and rear. The car was built by Harrold's PPJ racing team, with chief technician Trevor Chinn.*

**Left, below:** *The turbo installation in Harrold's rallycross Beetle is in front of the engine, above the gearbox, as seen in this view from inside the cockpit. The exhaust outlet is in front of the offside rear wheel.*

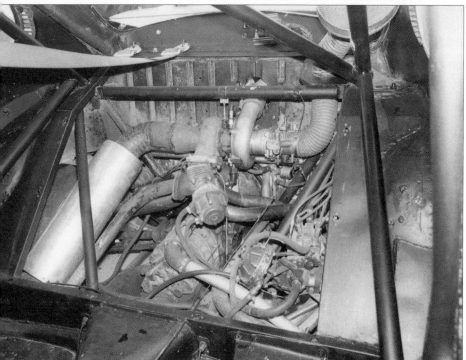

**Above:** *Away from its usual muddy habitat, the classic trials Notchback of Dean Vowden is seen on tarmac at the Curborough sprint course. This was a special test on the Chase Clouds Trial in 1990. Vowden won a first class award on the event; he also won practically everything there was to win in MCC classic trials in 1990 with this car. The Notchback is powered by a two-litre Type 4 engine with twin Webers. (A. Vowden)*

**Left:** *Mike Stephens won the BTRDA Production Car Trials Championship in 1991. Stephens' 1302S, built by Mike Hinde, has an SU carburettor, tuned for maximum torque at almost zero rpm, and runs with a massive amount of ballast bolted under the rear wings.*

124

## THE 1990s

*Above:* Graham Hoare became BTRDA All-Rounders Champion for the ninth time in 1992, driving Beetles in autotests and autocross and a Golf or Audi 80 in trials. The runner-up to Hoare, Charles Golding, used a Toyota in rallies and Ford Escorts in autocross and autotests.

*Above, right:* Hoare's autotest Beetle, a 1776cc 1303S, seen performing a reverse spin turn at a Maidstone & Mid-Kent Motor Club event. Hoare won the large saloon class of the BTRDA Autotests Championship in 1990, 1991 and 1992, the BTRDA Silver Star in 1990 and 1992, and he was a member of the victorious team representing England in the 1991 Ken Wharton Memorial Autotests.

*Right:* The interior of Graham Hoare's autotest 1303S is standard except for the 'twirling knob' on the steering wheel. This is essential for autotest manoeuvres to turn from lock to lock very rapidly with one hand, while the other hand changes gear (first to reverse) and operates the handbrake.

*Right:* Hoare's autocross Beetle is a 2.1-litre 1303S. Although primarily gaining points in the All-Rounders Championship, he succeeded in adding the 1992 BTRDA Autocross Class Championship to his other titles. In addition to his 11 autotest victories in 1992, Hoare scored six class wins in autocross and his Beetle is often in contention for outright fastest time. (Brooke Photographic)

# VW BEETLE IN MOTORSPORT

*Above:* Echoing VW successes in the 1950s in Round Australia Trials and African Safari Rallies, is Martin Garibay, from Mexico, who drives an almost-standard Beetle in desert races. Hardly any other production saloons can survive in these events. Garibay, here on his way to a class win on the Tecate/ SCORE Baja 500, is the High Desert Racing Association Class 11 Champion of 1991 and 1992. (BF Goodrich)

*Left:* Garibay's Beetle racing through the desert west of Las Vegas to record another class-winning performance on the Motorcraft Nevada 500 in 1992. Like most of the leading competitors in desert racing, Garibay's Beetle is equipped with BFGoodrich T/A tyres and Bilstein shock absorbers. (BFGoodrich)

*Left, below:* The Class 5 (1600cc) champion of desert racing is this Baja Beetle, built and driven by the Cook brothers, Wayne, Darryl and Alan, of California. The 'Cook'n VW' has a long list of victories including the Baja 500, Gold Coast 300, Nevada 500 and, as pictured here, the San Felipe 250. In this class engine modifications are very limited; single port heads and a stock Solex carburettor must be used. The chassis can be highly developed and the Cook brothers' car, originally a '67 Beetle, has a '61 linkpin front axle, special trailing arms and king-pins, rack and pinion steering, double-jointed rear suspension with Type 4 c.v. couplings, 181 rear drums, Type 2 transmission and twin Bilstein shocks. Suspension travel is more than 10 inches. Fuel capacity is 22 gallons. (BFGoodrich)

THE 1990s

*Above:* The unlimited Baja class in desert racing retains some 'VW style' components, but cars have a tubular space frame (with 105-inch wheelbase) and no roof. Type 4 engines are used and increased suspension travel is permitted. Seen here heading for a class win on the Baja 500 in 1992 is the BFGoodrich and Bilstein-equipped car of Hartmut and Wolfram Klawitter, who were the Class 5 Unlimited Champions in 1988, 1989 and 1991. (BFGoodrich)

*Right:* Mike Jakobson and Ron Jurkovac took their first desert race class win with this Type 4-engined unlimited Baja Bug on the 1992 Motorcraft Nevada 500. (BFGoodrich)

*Right:* 18 year-old Travis Howard drove his stock Beetle to victory in class 11 on the 1992 Fireworks 250 desert race in California. Howard started racing his Beetle in 1991, scoring wins on the Parker 400 in Arizona and the Gold Coast 300 in Nevada. He had a slight problem when the spare wheel became dislodged and jammed the bonnet open, but overcame it by driving for 10 miles looking through the glove-box aperture! (BFGoodrich)

*Left:* Jack Sacchette won the Pro Stock Championship three times with his 2.5-litre-engined Beetle, setting the ET record of 9.82 seconds. (K. Seume)

*Left:* Ron Townsend, seen here at Phoenix in 1991, is one of the top three in the Pro Turbo class and has achieved an ET of 9.4 seconds. (K. Seume)

*Below:* The Berg Family has been drag racing this Beetle since 1968 and claims it has won more races than any other VW, driven mostly by Gene's son, Gary. The car is used as a test bed for the tuning products sold by Gene Berg Enterprises. In 1991, when this photo was taken, the Beetle was running with Berg's newly-developed fuel injection system and posting under-10-second times. For 1992 it was converted back to carburettors to run in the Super Street class and immediately set a new class record at 10.55 seconds. (K. Seume)

## THE 1990s

*Right:* One of the most consistently successful drivers in the Käfer Cup is Dr Josef Gerold, who finished second in class and fourth overall in the series in 1989, 1990 and 1991 with his Type 4-engined 1302. Gerold has recorded the fastest Beetle lap times at Nürburgring, Hockenheim and Salzburgring. Here he is seen on the hillclimb at Unterfranken in 1991. (Gute Fahrt/Joachim Fischer)

*Right:* Ingo Gaupp leading Dr Jo Gerold and Bertram Krämer at Nürburgring. Gaupp was Division W4 Champion and second overall in the Käfer Cup in 1991, winning all the circuit races that year. His 1200 Beetle has a double-jointed rear axle, coil springs front and rear, Porsche Carrera brakes and 200bhp Type 4 engine. (Gute Fahrt/Joachim Fischer)

*Below:* Two of the stars of the Käfer Cup Division V (limited to a stock Solex 34 PICT carburettor), 'Fast Edy' Henckel (1500 Beetle) and Roger van Hoef (1303S), racing at Croix-en-Ternois, in northern France. 'Fast Edy' was the Class Champion in 1989. (Gute Fahrt/Joachim Fischer)

## THE 1990s

*Opposite page*

**Top:** *The Big Boys' Toys Beetle Cup brought all-Beetle racing to Britain for the first time in 1992. More than 20 Beetles, with very limited modifications, contested the ten-race series at Brands Hatch, Donington, Cadwell Park, Snetterton, Oulton Park, Castle Combe and Mallory Park circuits. Battling for mid-field placings here at Brands Hatch are BBC TV Top Gear presenter Chris Goffey (car no. 3) and The Very Wicked Car Company's Ray Purdham (no. 8). They are followed by Nick Mailer (21), Neil Birkitt (14) and Richard Smart (30).*

**Bottom:** *Julian Lock was unbeaten in the first nine Beetle races in Britain (four rounds of the Beetle Cup and five races at the Bug Prix). Although Simon Howarth scored a few wins towards the end of the season, Lock's claim to the Beetle Cup Championship title was never really in doubt. Lock also won the prestigious Bug Prix at Brands Hatch.*

**Right:** *Lock helping to sort out a problem with the carburettor on John Aitkenhead's Beetle before the race at Donington.*

**Below:** *Close racing in the Bug Prix, involving Grant Cassidy (15), Jim Gray (2) and Autosport's Marcus Pye (99).*

VW BEETLE IN MOTORSPORT

*Above:* An electrical problem in Tony Cook's Beetle receives attention at Brands Hatch.

*Left:* Peter Nicholas checking his Beetle's valve clearances before the race at Cadwell Park.

*Above:* Beetle Cup racing became closer as the season progressed. This was the penultimate race of 1992, at a mist-shrouded Donington Park. Attempting to negotiate Goddard's Bend simultaneously are Les Goble (18), Colin Wells (44), Simon Miles (1), Peter Keys (5) and Jason Winter (4). Goble finished third in this race, behind Howarth and Miles, after Julian Lock was slowed by loss of power due to carburettor icing. Keys and Winter achieved top five placings in the Championship.

*Right:* Dave Butler discovering traction in his 1776cc supercharged Beetle which reached 123mph at Australia's Willowbank Raceway, covering the quarter-mile in 11.4 seconds. The Vintage Vee-Dub Supplies-sponsored Beetle is fitted with aerodynamically-designed mudguards. (R. Hölzl)

VW BEETLE IN MOTORSPORT

**Left:** Richard Hölzl is one of several Australians who race Beetles against other makes in the road-registered and sports sedan classes of circuit racing. Hölzl's best result is third overall and he nearly always finishes in the top six in races of more than 40 cars, including General Motors V8s, Mazda RX-7s, Ford Escort Cosworths and Alfa Romeo V6s. Hölzl's Beetle has also won a drag race, beating a Ford Falcon GT V8 in front of 20,000 spectators, and he still regularly drives it on the road! (R. Hölzl)

**Below:** Hölzl's oval-window Beetle on a rolling road. Power at the wheels is 140bhp at 5000rpm. The engine is 2330cc with Weber 48 IDA carbs. The 1956 oval body is on a '68 chassis with double-jointed rear suspension and Porsche 914 rear disc brakes. The car was built with much help from Hölzl's friend and mentor, Henry Spicak. (R. Hölzl)

**Opposite page**

**Top:** Don and Kurt Meyer race this Beetle in the VW Sedan Class at circuit races organised by the Midwestern Council of Sports Car Clubs. The rules limit modifications, requiring a standard-size carburettor and stock valve sizes, but permit under-1500cc Golfs and Sciroccos in the same class as Beetles. Pictured at the Blackhawk Farm Circuit near Beloit, Illinois, Meyer's '69-model Beetle has a 1584cc engine (the maximum allowed) with Solex 34 PICT carb, modified camshaft, gasflowed heads, high compression, and Berg sump. The chassis has a stiffened and lowered double-jointed rear suspension and disc brakes. The Americans have regarded swing-axle suspension and drum brakes as totally unsuitable for circuit racing since Art Schmidt rolled his old swing-axle car in 1983. (K. Meyer)

**Bottom:** Four Beetles racing at the Road America Circuit, Elkhart Lake, in 1991: Kurt Meyer, Art Schmidt, Jim Uppenkamp and Ron Petrucci. The VW class races with other makes such as MG Midget, Triumph Spitfire, Saab, Toyota, Porsche 914 and small-engined Mustangs. (K. Meyer)

*THE 1990s*

## THE 1990s

*Opposite page*

**Top:** Three generations of the Davis family, from Buckinghamshire, have driven VWs in trials. Jack (left of picture), now 88 years old, was an active competitor until the mid-1960s. He is still involved in motorsport as vice-president of the MCC and now drives a '73 model 1303 Beetle — although no longer in competition! His son Michael (right) and grandson Ian (centre) compete in trials, driving both the 1300 Beetle seen here and a trials Buggy previously owned by Richard Penhale. (M. Davis)

**Bottom:** Ian Davis, with grandfather Jack, in the Buggy at Brooklands on the occasion of the MCC's 90th Anniversary Trial in 1991. (M. Davis)

**Right:** Bob Beales driving his '58 Okrasa Beetle on the Charrington's RAC Historic Rally in 1992. Beales won the Welsh Championship with this car, but had trouble on the Charringtons; a broken coupling in the gear linkage and a bent stub axle, due to an encounter with a kerbstone, caused delays and he finished fifth in class.

**Below:** The interior of Beale's rally Beetle is functional and well-used. Not surprisingly, he lost out on one rally when he tied for first place and the tie-breaker was a concours d'elegance!

## VW BEETLE IN MOTORSPORT

**Left:** A 1957 Beetle, driven by Belgians Jean-Jacques Faignart and Michel Dartevelle, with sponsorship from Ethyl Petroleum Additives, on the Liège-Rome-Liège Historic Rally in 1991. The Beetle crew were lying fourth until a navigational error dropped them off the leader board. The rally's 4,000-mile route crossed the Alps and the Dolomites. (Ethyl)

**Below:** Helen Marin received this beautiful 1955 oval-window Beetle as a birthday present! Powered by a Terry's Beetle Services-built 1835cc engine, with twin Dellortos, the Oval has run a best quarter-mile time of 15.8 seconds and Helen is a regular competitor in the Street class at VWDRC events. She drives it on the road every day, even using it to take her disabled mother to Bingo! And with immaculate black paintwork, highlighted at the front by green flame-effect graphics, and fitted with Scirocco seats, the car has also won show'n'shine trophies.

**Bottom:** Another contender in the Street class is Dafydd Charles, who lets everyone know he is Welsh by acknowledging his crew with the wording 'Da di'r Hogia Jimmy & Ger' (roughly, 'the lads have done well, Jimmy and Geraint') on his Beetle's engine lid. This VW started its motorsport career in autotests when David Jones drove it as a member of the Welsh team in the Ken Wharton event. It also did some rallying before Charles acquired it. He rebuilt the car, fitting a 2.1-litre Type 4 engine with twin Webers, close-ratio gears and limited slip diff. Best quarter-mile time for this good-looking, daily-driven Beetle is 15.4 seconds.

### Opposite page

'Bad Bernie' Smith, one of the originators of VW drag racing in Britain, returned to the strips in 1992 with this awesome new dragster, powered by a 2.8-litre, 600+bhp, turbocharged, Autocraft VW-style engine, and immediately challenged Paul Hughes for the 'Fastest European VW' title with low-eight-second 160mph runs.

VW BEETLE IN MOTORSPORT

*Left:* The Kits and Buggies Class Champion for 1992 was this 1911cc Type 4-engined Charger, driven by Mark Lefley. The Charger runs mid-15 second times. Cars in this class must be street legal.

*Below:* If you wonder why Paul Sutton calls his Beetle 'Yum Box', just look at the registration number! This Volksline of Canterbury and John Maher-supported Super Street entry has a 2.3-litre engine with nitrous. Sutton ran mid-12-second times in 1992 and finished the season third behind Bowen and Angell.

*Bottom:* Malcolm Palmer runs an Autocavan agency in Worcester called 'Bug Bits' and this is his Super Street Beetle, called 'Bugsy Malone'. With a 1600cc engine, and still in use as a daily driver, its best time was a very good 15.75 seconds. This shot was taken on the car's first run with a 2165cc engine, with twin Dellorto 48 carbs (no nitrous, yet) and close ratio gears. Palmer expects to be doing 12-second runs in 1993.

**Opposite page**

*Top:* Brian Burrows checks the rear tyre pressure of his Beetle Funny Car drag racer 'Outrage 2', which in 1992 was the fastest and most exciting flat-four-engined car in Europe. Burrows, a graphic designer, has given the driving seat in his 160mph 'Beetle' to Paul Hughes, but his unbounded enthusiasm for VWs and drag racing has led him also to promote VW events at Avon Park, where he has done much to improve the image and popularity of the sport and to reduce costs for those who attend.

*Bottom:* Paul Hughes launches 'Outrage 2' on an eight-second run. Mechanical details include a turbocharged Autocraft VW-style engine, in front of the driver, a Chevrolet Powerglide transmission and Ford solid back axle. The chrome-moly tube chassis was designed and built by Bob Nixon (in Essex) and the lightweight fibreglass body came from Creative Car Craft (in Florida). Retardation is provided by two Simpson parachutes. VW Dream Machine provides sponsorship for the project.

## THE 1990s

## VOLKSWAGEN OWNERS CLUB of Great Britain

Established 1953

The Volkswagen Owners Club of Great Britain was founded in 1953 and has grown to be the largest organisation of its kind for VW enthusiasts in this country. Benefits enjoyed by members are what one would expect from an organisation that has catered for the enthusiast for 40 years. If you wish to cut your motoring costs then we can help with discounts on spares, insurance and agreed value insurance, RAC membership with associated member status, books and magazines to name but a few. We can - of course - supply the technical advice on how to restore, repair or find that elusive part for your vehicle. If you wish to participate in socials or competitions we are the experts both nationally and regionally. We organise the Clee Hill Trial - voted more than once " the best organised trial in the country " as well as organising the VW Audi National Autotest and we assist each year with the running of the British Volkswagen Festival. Our members receive our club magazine ClubNews on a bi - monthly basis as well as regional newsletters, our discount directory and our Mutual Aid Register. Want to know more? Yes ? then send a SAE to
THE MEMBERSHIP SECRETARY VWOC (GB),
PO BOX 7, BURNTWOOD,
WALSALL, WS7 8SB
for full details re membership, subscription rates etc.

## LONDON LEISURE LINES
8B ALEXANDRA PARK ROAD
LONDON N10 2EG
Tel: 081-883 8688 . Mobile: 0831-616501 . Fax: 081-442 1476

### EVERY CHILD'S DREAM

Children's V.W Beetle Cabriolet. Safe, Quiet and Maintenance Free.
Floor gear shift • 2 Forward gears and reverse • Pedal accelerator • drum brakes • Working headlights rear lights • Shock absorbers and suspension • Rack and pinion steering • Opening Bonnet with spare wheel • Opening engine lid with engine • Opening doors • Twin 240W motors • Twin batteries with mains charger • Wing mirrors • Cassette player with small casette • Seat belts • 2 Ignition keys • Glove box • Driving licence and log book • V.W. licenced • Highest quality, 12 months guarantee (Parts and labour) to EEC and BS safety standard. Complete - £545 including delivery.
AS SUPPLIED TO V.A.G. DEALERS NATIONWIDE

WE AT SIGNCRAFT CAN TRANSFORM YOUR CAR WITHOUT MAJOR OUTLAY. FROM A SINGLE COACHLINE TO A FULL BLOWN MULTI DESIGN, YOUR IMAGINATION IS THE LIMIT! FOR A DESIGN WHICH STANDS OUT PROUD CALL US NOW !

Thame (084421) 3842  **Signcraft**

VEHICLE GRAPHICS — VINYL SIGNWRITING — DESIGN SERVICE — VISUALISING — EXHIBITIONS — CORPORATE ID — MARINE & AVIATION — COMMERCIAL FACIAS

## FRANCIS TUTHILL'S WORKSHOP

Success on London to Sydney 1977,
RAC Rally three times,
plus many other competitions combined with 20 years experience makes us very able to help you with any Beetle requirement.

Service, Bodywork, Restoration, Roll Cages are all part of our range.

*Please call us for more details.*

THE WORKSHOP, WARDINGTON, NR BANBURY, OX17 1RY
TELEPHONE:
0295 750514

## OXFORD BEETLES

DAYTIME 0235 770996   24HR A/PHONE 0235 770926

Restoration & Repairs
Customising, Servicing & Spares
Sales & To Order to Your Own Spec

ANY CAR MoT'd — CAR GUARANTEED — VOLKSWAGEN — QUALITY — SUSPENSION & ENGINE MOD's — VAN

UNIT ONE, STATION YARD, GROVE, WANTAGE, OXON

It's all your Beetle needs ★ Special student rates

## CONGLETON BEETLES

*For VW Beetle Restoration, Sales, Servicing, Parts & Repairs. Bodywork a Speciality. Resprays. Cal Look and Custom Work Undertaken. All Work Guaranteed.*

Danemill, Broadhurst Lane,
Congleton, Cheshire.
Tel: 0260 - 279887

**SAVE MONEY**
Repair your own VW-Audi

We sell nearly 200 specialist VW tools. Comprehensive mail order catalogue £3. Refundable with purchase.

**338 Bradford Road, Liversedge, West Yorkshire WF15 6BY
Telephone: 0924 402860**

**John Forbes Automotive**
7 Meadow Lane
Edinburgh
EH8 9NR
**031 667 9767**

**OVER 30 YEARS BY THE BEETLEMASTER IN VW's**

Spares – Service – Repairs
**Autocavan Agent**

'We have the key to keep your Beetle jumping!'

**VOLKSWORK**
Beetle Specialists

Servicing • Brakes • Suspension
Clutches • Bodywork • Welding
Grit-blast and Bead-blast Facilities
Roll-bars and cages made and fitted

Unit 6P, Atlas Business Centre,
Oxgate Lane, London NW2 7HJ

Contact Chris or John, 081 450 1004

# Appendix

*For those who would like to drive their Beetles in motorsport...*

*VW Motoring*

The cheapest forms of motorsport are trials, autotests, some types of road rallies and the Street class of drag racing. For these it is not necessary to have protective clothing, crash helmets or roll cages. For all except drag racing it is necessary to be a member of a motor club recognised by the RAC Motor Sports Association and, for all except the lowest level of events, an RAC competition licence is required.

Standard cars can be used for the above-mentioned events but the following basic advice should be borne in mind:

- ◆ Adequate ground clearance is essential for trials. The car should certainly not be lowered; most successful VW trials cars have raised suspension. Low-mounted spoilers, foglamps, exhaust systems etc. will risk being damaged. The regulations must be checked regarding permitted wheel and tyre size and tyre type.
- ◆ Autotests involve performing tight turns where oversteer is an advantage. Modifications to improve a VW's handling on the road, e.g. decambering and wider rear wheels, will be a hindrance rather than a help in autotests, although lowered suspension is desirable to prevent the risk of overturning.

Autotests and trials both depend far more on driving ability and technique, and far less on car specification and power, than any other type of motorsport. Both are perfectly suitable for novices, but overnight success is unlikely. In autotests the driver must be able to memorise the often complex routes.

For Beetle Cup racing a racing licence must be obtained from the RAC Motor Sports Association; a prerequisite for the licence is to attend a course (usually one day) at an approved race driving school. For racing, autocross, rallycross and special stage rallies, safety equipment including roll cage, fire extinguisher, battery master switch, laminated windscreen, crash helmet and fire-resistant clothing is compulsory.

In both the Beetle Cup and historic rallying the degree of modification is closely controlled; careful study of the regulations is essential.

Rallycross and American desert racing are the roughest and toughest forms of motorsport; these events are definitely not for the novice or the faint-hearted!

VW BEETLE IN MOTORSPORT

*VW Motoring*

## Addresses

**Association of Classic Trials Clubs.**
A. Templeton, The Coach House, Chivers Road, Stondon Massey, Brentwood, Essex CM15 0LG.

**Autotest Drivers' Club of Northern Ireland.** D.J. Coates, 11 Alandale, Bangor, Co. Down, N. Ireland BT19 2DF.

**Big Boys' Toys Promotions.**
Unit 1, Motherwell Way, West Thurrock, Grays, Essex RM16 1NR.

**British Racing and Sports Car Club.**
J. Sherwood, Brands Hatch, Fawkham, Dartford, Kent DA3 8NH.

**British Trial and Rally Drivers' Association.** Liz Cox, 19A Oxford Street, Lambourn, Berks RG16 7XS.

**Club VW Sydney.** P.O. Box 1135, Parramatta, NSW 2124, Australia.

**High Desert Drag Racing Association.**
12997 Las Vegas Boulevard, Las Vegas, NV 89124, USA.

**Historic Rally Car Register.**
Ms A. Woolley, Tibberton Court, Tibberton, Gloucester GL19 3AF.

**International Volkswagen Association (IVWA).** PO Box 25123, Winston-Salem, NC 27114-5123, USA.

**Käfer Motorsport.**
Klaus Morhammer, P-1548, 8035 Gauting, Germany.

**Midwestern Council of Sports Car Clubs** (VW Sedan Racing). K & D Racing, 302 Chamberlain Avenue, Madison, WI 53705, USA.

**Motor Cycling Club.** G.M. Margetts, Hares Bank, 21 Macclesfield Road, Malvern, Worcs WR14 2AS.

**RAC Motor Sports Association.**
Motor Sports House, Riverside Park, Colnbrook, Slough SL3 0HG.

**SCORE International**
(Short course off-road racing),
31125 Via Colinas, Suite 908, Westlake Village, CA 91362, USA.

**Texas VW Drag Racing Association.**
1047 Cedar Run, Duncanville, TX 75137, USA.

**VW Drag Racing Club.**
Gwynn Thomas,
15 St Cross Road, Crondall, Farnham, Surrey GU10 5PQ.

**Volkswagen Owners Club (GB).**
Simon Holloway, 71 Hamble Drive, Abingdon, Oxon OX14 3TF.

# Autocavan

## THE AIR-COOLED LEADER SINCE 1968!

## 1968 TO 1993

## 25 YEARS

**AUTOCAVAN HEAD OFFICE:** 103 Lower Weybourne Lane, Badshot Lea, Farnham, Surrey
Tel: (0252) 27627/333891  Fax: (0252) 343363  Open Mon – Fri: 8.30 – 5.30 Sat: 8.30 – 4.00

| | | | | | | | |
|---|---|---|---|---|---|---|---|
| **Autocavan Rochdale:** | 0706 526286/7 | **Autocavan Belfast:** | 0232 772535 | **Autocavan Exeter:** | 0392 52075/211414 | | |
| **Autocavan Ipswich** | 0473 461341/748166 | **Autocavan Poole:** | 0202 722483/722535 | **Autocavan Glasgow:** | 03552 35666 | | |

**WITH 49 AUTOCAVAN APPROVED DISTRIBUTORS, OUR QUALITY PARTS ARE AVAILABLE ALL OVER THE UK — THERE'S ONE NEAR YOU!**

| | | | | | | | | | |
|---|---|---|---|---|---|---|---|---|---|
| Abergavenny: | 0873 3774 | Canterbury: | 0227 767958 | Kidderminster: | 0562 752497 | Redcar: | 0642 475631 | Warminster: | 0985 40153 |
| Ashstead: | 03722 77888 | Castleford: | 0977 518254 | Kilmarnock: | 0563 71050 | Redruth: | 0209 217161 | Warrington: | 0925 821211/810541 |
| Ashton-under-Lyne: | 061 343 2487 | Chelmsford: | 0245 359960 | Leamington Spa: | 0926 451980 | Rhyl: | 0745 590272 | Warwick: | 0926 400516 |
| Bedford: | 0234 354237 | Cheltenham: | 0452 863335 | London (Plumstead): | 081 317 7333 | Sheffield: | 0742 730606 | Wellingborough: | 0933 227277 |
| Blackburn: | 0254 57907 | Chorlton (Manchester): | 061 881 5225 | Mildenhall: | 0638 712519 | Shrewsbury: | 0743 241418 | Wisbech: | 0945 64650 |
| Blyth: | 0670 353131 | Eastbourne: | 0323 21827 | Nottingham: | 0602 503723 | Slough: | 06286 67464 | Worcester: | 0905 724888 |
| Bristol: | 0272 717057/717079 | Edinburgh: | 031 667 9767 | Oldham: | 061 678 9610 | Southampton: | 0703 787118 | Worsley: | 061 790 5483 |
| Burton-on-Trent: | 0283 33337 | Eire (Tipperary): | 010 353 62 76177 | Plymouth: | 0752 227116 | Stockport: | 061 483 3883 | Worthing: | 0903 783373 |
| Caernarfon: | 0286 3559 | Eire (Wicklow): | 010 353 404 67189 | Pocklington (Nr York): | 0759 303716 | Stoke-on-Trent: | 0782 834804 | Yeovil: | 0935 77312 |
| Cambridge: | 0223 425600 | Hereford: | 0432 277046 | Portsmouth: | 0705 824028 | Street (Somerset): | 0458 42707 | | |

## 25 YEARS OF COMPETITION SUCCESS!